# NATURE'S PATTERNS

Philip Ball is a science writer, and the author of many popular science books including *H₂O: A Biography of Water*, *Bright Earth*, *Critical Mass* (winner of the 2005 Aventis Prize for Science Books) and *The Music Instinct*. He lectures widely and has contributed to magazines and newspapers, including *Nature*, *New Scientist*, *The Guardian*, and *The New York Times*.

# NATURE'S PATTERNS

## A Tapestry in Three Parts

### PHILIP BALL

Nature's Patterns is a trilogy composed of
Shapes, Flow, and Branches

OXFORD
UNIVERSITY PRESS

# OXFORD
UNIVERSITY PRESS

Great Clarendon Street, Oxford OX2 6DP

Oxford University Press is a department of the University of Oxford.
It furthers the University's objective of excellence in research, scholarship,
and education by publishing worldwide in

Oxford New York

Auckland Cape Town Dar es Salaam Hong Kong Karachi
Kuala Lumpur Madrid Melbourne Mexico City Nairobi
New Delhi Shanghai Taipei Toronto

With offices in

Argentina Austria Brazil Chile Czech Republic France Greece
Guatemala Hungary Italy Japan Poland Portugal Singapore
South Korea Switzerland Thailand Turkey Ukraine Vietnam

Oxford is a registered trade mark of Oxford University Press
in the UK and in certain other countries

Published in the United States
by Oxford University Press Inc., New York

© Philip Ball 2009

British Library Cataloguing in Publication Data

Data available

Library of Congress Cataloging in Publication Data

Data available

Typeset by SPI Publisher Services, Pondicherry, India
Printed in Great Britain
on acid-free paper by
Clays Ltd., St Ives plc

ISBN 978-0-19-960486-9

1 3 5 7 9 10 8 6 4 2

# SHAPES

There is beauty to be found in regularity: the same element repeating again and again, typically with geometric order. There is no better example than the honeycomb, a miracle of hexagonal perfection. This sort of pattern is a *lattice*. On a leopard's pelt, the lattice melts but a pattern remains: spots spaced at more or less equal distances but no longer in neat rows. There is a comparable order in stripes and concentric bands: a succession rather strictly enforced on the angelfish, but more loosely applied in the meandering, merging stripes of the zebra or of sand ripples. The means by which these natural patterns are constructed may tell us something about how the far more complicated forms of animals and plants are created by a progressive division and subdivision of space, orchestrated by nothing more than simple physical forces.

# CONTENTS

# PREFACE AND ACKNOWLEDGEMENTS

After my 1999 book *The Self-Made Tapestry: Pattern Formation in Nature* went out of print, I'd often be contacted by would-be readers asking where they could get hold of a copy. That was how I discovered that copies were changing hands in the used-book market for considerably more than the original cover price. While that was gratifying in its way, I would far rather see the material accessible to anyone who wanted it. So I approached Latha Menon at Oxford University Press to ask about a reprinting. But Latha had something more substantial in mind, and that is how this new trilogy came into being. Quite rightly, Latha perceived that the original *Tapestry* was neither conceived nor packaged to the best advantage of the material. I hope this format does it more justice.

The suggestion of partitioning the material between three volumes sounded challenging at first, but once I saw how it might be done, I realized that this offered a structure that could bring more thematic organization to the topic. Each volume is self-contained and does not depend on one having read the others, although there is inevitably some cross-referencing. Anyone who has seen *The Self-Made Tapestry* will find some familiar things here, but also plenty that is new. In adding that material, I have benefited from the great generosity of many scientists who have given images, reprints and suggestions. I am particularly grateful to Sean Carroll, Iain Couzin, and Andrea Rinaldo for critical readings of some of the new text. Latha set me more work than I'd perhaps anticipated, but I remain deeply indebted to her for her vision of what these books might become, and her encouragement in making that happen.

Philip Ball

*London, October 2007*

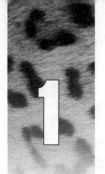

# THE SHAPES OF THINGS

## *Pattern and Form*

rriving on Earth, the aliens approach the first thing they see and
utter the familiar words: 'take me to your leader' (Fig. 1.1). Like
many jokes, this one offers a damning critique. It undermines the
venerable and serious scientific quest to find life on other worlds, explod-
ing the question of 'how would we *know* if we found it?' by answering that
we tend to imagine it will look like us.

Now, let me assure you that astrobiologists (as scientists who study
aliens are called nowadays) are not really that foolish. They do not
imagine for a moment that when we touch down on another inhabited
world, we will be greeted by envoys who look like Leonard Nimoy.
Indeed, if there is life in those parts of our own solar system that seem

FIG. 1.1   Do we inevitably expect life to 'look' like us?

I

at all habitable (such as the subsurface seas of Jupiter's icy moon Europa), it is most unlikely to warrant the description 'intelligent'. And we may have to look hard and long to find it, precisely because we don't know what we're looking for. Yet even if we know it is not going to be Dr Spock, we have trouble shaking the conviction that it will look something like the forms of life we have seen before.

Even if it did, that already makes the challenge of identifying extraterrestrial life bad enough. Take a look at life on Earth today, and you'll see such a bewildering variety of shape and form that you could be forgiven for imagining anything is possible (Fig. 1.2). But scientists have a rather more sophisticated view of life (although they still cannot agree on a universal definition of it), which gives them hope of distinguishing it from its inorganic context. They recognize some attributes of living systems that go beyond mere physical appearance, such as the fact that life tends to destroy the chemical equilibrium of its environment. I'll explain later what I mean by that, but let's say for now that it's rather like watching a film in which all you can see are balls being juggled: you know that there is something out of frame that is keeping them in motion. It's true that some geological and astrophysical processes that don't involve life at all can also induce this disequilibrium—but nonetheless, searching for disequilibrium as a potential fingerprint of life seems a lot better than looking around for a loitering humanoid alien to whom you can say 'Take me to your leader'.

Nevertheless, old habits die hard. Meteorite ALH84001 is a potato-shaped lump of Mars that was blasted from the Red Planet a few billion years ago in an asteroid or meteorite impact and subsequently found its way through space to Earth. It was discovered in 1984 in the snows of Antarctica. Scientists who made a detailed study of this cosmic intruder claimed in 1996 that it contains a 'possible relic' of Martian life. In support of that claim, an image was broadcast around the world that seemed to show worms crawling across the mineral surface (Fig. 1.3). These 'worms' were mineral too, and so tiny that they could only be seen in the electron microscope; but the suggestion was that they could be the fossilized remains of Martian bacteria that once infested this chunk of stone.

The researchers who investigated ALH84001 admitted that this conclusion was tentative, and they didn't make it lightly. These wormy forms

FIG. 1.2  Living organisms on Earth come in a bewildering variety of shapes and sizes. (Photos: a, carolsgalaxy; b, Keenan Pepper; c, Sarah Nichols; d, twoblueday; e, Ed Schipul; f, Doug Bowman.)

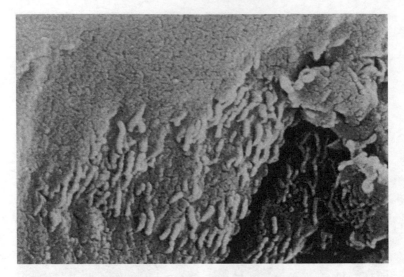

FIG. 1.3 These microscopic structures found in Martian meteorite ALH84001 have been interpreted as evidence of ancient bacterial life. Might they be the fossilized remnants of tiny organisms? (Photo: NASA.)

were by no means the sole evidence—and after all, the scientists acknowledged, they were much smaller than earthly bacteria tend to be. All the same, these structures didn't look like inorganic forms: it was hard to explain them as microscopic rock features formed by physical forces alone. And so the researchers stuck out their necks and used shape, pattern, form—what scientists tend to call *morphology*—as the partial basis for inferring a possible signature of life.

That doesn't seem an unreasonable thing to do, does it? Surely, after all, we can distinguish a crystal from a living creature, an insect from a rock?

Well, you might think so. But take a look at Fig. 1.4. At the top are the shells of marine creatures called diatoms (which we will meet again shortly). Below are microscopic mineral formations created in a test tube, entirely without the agency of life. Would you trust yourself to say which is a 'living' form, and which is not? Now look at Fig. 1.5, in which the microscopic patterns are a product of much the same chemical process that made those in 1.4b. Does this remind you of anything?

4

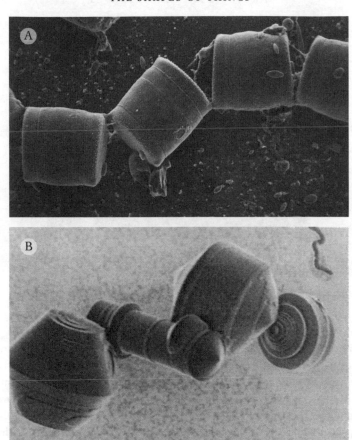

FIG. 1.4 Can we tell apart inorganic structures formed by chemistry alone and those wrought by biology? The shapes in *a* are the shells of marine micro-organisms called diatoms; but while those in *b* have a similar complexity that seems also to speak of the agency of living organisms, they are the product of a purely chemical process. (Photos: *a*, Rex Lowe, Bowling Green State University, Ohio. *b*, Geoffrey Ozin, University of Toronto.)

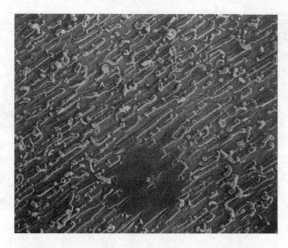

FIG. 1.5 Chemistry alone can also produce microscopic surface patterns rather similar to those seen in meteorite ALH84001 (Fig. 1.3). (Photo: Geoffrey Ozin, University of Toronto.)

What is it that encourages us to typecast some forms as those made by life, and others as the products of the non-living world? A tree, a rabbit, a spider have rather little in common when considered as mere shapes—and yet we don't hesitate to see them as examples of living morphology. Why? Perhaps we sense a kind of purpose, of design, in these forms? They are 'complex', certainly, but what does that mean? They may have some regularity or symmetry—the bilateral symmetry of the animals, the repeated branching of the tree—but that can't be all there is to it, for surely it is often in a high degree of regularity and symmetry, in the onset of crystallinity, that we might imagine we discern life's absence. Even if we can't say exactly what living form is, we'd like to think we know it when we see it. But do we?

In the late 1990s, a group of NASA researchers decided this was a problem that, like many others, is best left to a computer. They believed that artificial intelligence would be more likely than we are to distinguish the living from the non-living, and so they hoped to 'train' computers to recognize life from morphology alone, using all the examples they could muster from our own planet. The machine would sense and assimilate the subtle characteristics of living form, and would then look for these signatures on missions to Mars or other potentially life-supporting worlds. They intended this to be another of the computationally intensive projects like that which analyses radio signals for signs of extraterrestrial intelligence,* in which volunteers' home computers analyse the data—in this case, running the computationally intensive learning process—during spare time. This distributed system would feed its output into the central brain of the operation, a computer that the NASA researchers proposed to call the D'Arcy Machine.

Now that seems truly ironic, for the name is inspired by the man who, to my mind, did more than anyone else to undermine the notion that life has characteristic forms that distinguish it from non-living systems. This man was the Scottish zoologist D'Arcy Wentworth Thompson, whose classic book *On Growth and Form*, first published in 1917, provides the first formal analysis of pattern and form in nature. Thompson's book was deeply erudite, beautifully written, and like nothing that had gone before. It was also several decades ahead of its time—and this, coupled with its undoubted idiosyncrasy, has meant that *On Growth and Form* never sparked the emergence of 'morphology' as a scientific discipline. For a long time, no one really knew what to make of Thompson's tome.

*On Growth and Form* presents a challenge to the naive view that non-living systems can produce only 'simple', often classically geometric, shapes and forms—the prismatic shapes of crystals, say, or the sterile ellipses of planetary orbits. Physics apparently teaches us that the basic laws of nature are simple and symmetrical, and so it seems natural to expect that their manifestations will share that characteristic. By the same token, we tend to imagine that life is a complicated and infinitely plastic influence, generating complex shapes that geometry struggles to encompass or describe.

D'Arcy Thompson argued that, on the contrary, life's forms may often echo those of the inorganic world, and both of them may be rather simple, or conversely, delightfully subtle and complex. Many of them, moreover, have an undeniable beauty, whether that be the elegant, Platonic beauty that we seem to perceive in symmetry and regularity or the dynamic, organic beauty of living nature. The ambitions of the D'Arcy

*This project, called SETI@Home, is an example of so-called distributed computing.

Machine notwithstanding, life leaves no characteristic signature in natural form.

What makes Thompson's book especially appealing is that there is nothing at all recondite about the things he discusses. These forms are ones we see all around us: the spirals of snail shells or in a sunflower head, the baroque whorls of flowing water, the lacework of clouds, the stripes of a zebra, the orderly perfection of a honeycomb. In these three books I shall consider many other examples which Thompson barely touched on, if at all, such as the designs of a butterfly wing, the undulations of sand dunes, the branching of trees and rivers. Science now possesses the tools and concepts to unravel the processes that create these things, and it has vindicated Thompson's approach of searching for universal physical causes of pattern and order, offering us a glimpse of a kind of natural harmony that pervades and structures the whole world.

The puzzle of how these structures arise has solutions that are both surprising and inspiring, and that foil our intuitions about how complex form and pattern are made. Many of the most striking examples that we encounter around us are evidently the products of human hands and minds—they are patterns shaped with intelligence and purpose, constructed by *design*. Brickwork tiling schemes, the horizon-spanning stepped terraces of Asian rice fields, the monotonous regularities of the built environment, the delicate traceries of microelectronic circuitry (Fig. 1.6)—all bear the mark of their human makers. The unconscious message that we take away from all this artifice is that patterning the world, shaping it into forms that please us or do useful things, is hard work. It requires dedicated effort and skill. Each piece of the picture must be painstakingly put into place, whether by us or, in the wider world, by the forces of nature: the sparrow building its nest, the individual plants weaving together into a hedgerow. This, we have come to believe, is the way to create any complex form.

So when the natural philosophers of ages past found complexity in nature, it is scarcely surprising that many of them decided they were gazing on God's handiwork and artistry. Most famously, the English pastor William Paley argued in his book *Natural Theology* that the contrivances found in the living world, in the forms of animals and plants, are so exquisite that they cannot but be the products of a guiding intelligence.

8

FIG. 1.6 Most of the complex patterns that we create are the products of painstaking labour, each element of the pattern having been put in position 'by hand.' Here I show a wall pattern from a Korean palace (*a*), a paddy field in China (*b*), an apartment block in Benidorm (*c*) and circuitry on a microprocessor chip (*d*). (Photos: *a*, Craig Nagy; *b*, McKay Savage; *c*, Ross Goodman; *d*, Digital Equipment Corporation.)

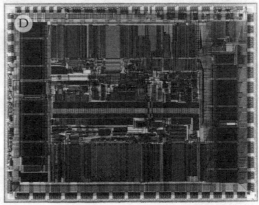

The shabby tatters of this idea survive today in the Intelligent Design movement, but Paley's argument was in its time much more defensible and indeed more coherent than that. For it was not until the latter half of the nineteenth century that Darwin's theory of evolution by random mutation and natural selection furnished an explanation of how apparent 'design' might arise in nature without a designer.

Contrary to the usual perception, Darwinism did not wholly answer Paley, for the reverend also considered that the principles of astronomy show signs of God's wisdom, for example in the physical layout of the solar system and the stability, simplicity and ponderous pace of planetary orbits. His arguments here are not terribly convincing, although it has to be acknowledged that modern cosmology is now even more of a happy hunting ground for those who hope to discern evidence of God's design. But Paley could have availed himself of many suggestive examples of 'design' in nature that did not come from the living world, and it is rather surprising that he did not. Anyone looking at the world with Paley's convictions about a Divine Architect would surely have to conclude that He possesses an irrepressible urge to create beauty for beauty's sake. Why else, for instance, is the snowflake—a mere ice crystal—wrought with such extravagance? Why are clouds dashed and dotted across the sky in orderly ranks? And can it be mere coincidence that we see the same forms and patterns again and again in places that apparently have nothing in common, not even the characteristic of being alive? Why should river networks resemble our veins and arteries, or whirlpools look like galaxies, if these are not the motifs chosen by God?

It is unfair, I know, to create a new argument for Paley only with the intention of demolishing it. But that is not really my aim. To my mind, the fact that these patterns can arise, and that they share certain elements, quite without any need for a Great Patterner, is much more remarkable and exciting than the notion that they are merely the products of a cosmic artisan weaving nature's tapestry. For that is just what happens. We would not expect to be able to make patterns such as those in Fig. 1.6 by, say, letting dyes spontaneously unmix into labyrinthine patterns, or trusting soil and pebbles to arrange themselves into terraces ready to take water and seed. But equivalent things really do happen, and in that way nature weaves its own designs. And what is more, we find that some of nature's patterns recur in quite different situations, as though this tapestry is woven from an archetypal design book, its themes echoing one another in different parts of the fabric. We shall see that nature's artistry is spontaneous, but is not arbitrary.

## THE BOUNDARIES OF EVOLUTION

The quixotic flavour of *On Growth and Form* stems not only from the fact that Thompson was attempting to explain many things for which he lacked the right instruments. He was also taking a stand against what he considered to be a stifling orthodoxy imposed by those who, at the beginning of the twentieth century, sought explanations for form and pattern in the living world. For although Darwinism obviated the need for Paley's natural theology, it seemed at times in danger of conducting the same conjuring trick by another name. Instead of proclaiming 'God's work' in response to every example of apparent design in nature, there was a tendency to exclaim 'evolution's work!' instead.

Darwin argued that, given enough time, small random changes in the forms of organisms could carry them towards those shapes that were best adapted to the demands that their environment made on them, because the struggle for survival weeded out those changes that made survival harder while conferring a reproductive advantage on those individuals who, by sheer luck, acquired a beneficial mutation.

This means that, in biology, it is natural to expect that form follows function: that the shape and structure of a biological entity (which could be a molecule, a limb, an organism, even a whole colony) is that which best equips the associated organism for survival. Biologists are still divided about whether the selective pressure that dictates such forms acts primarily at the level of the individual genes responsible for that characteristic, or of the whole organism. But either way, the implication is that form is *naturally selected* from a palette of possibilities. A form that confers evolutionary advantage tends to stick.

This is an extraordinarily powerful idea, and no serious biologist doubts that it is basically correct in explaining how organisms evolve and adapt over time. But as an explanation for the forms of life, it is not entirely satisfying—not because it is wrong, but because it says nothing about proximate mechanism, about causes that operate not over evolutionary time but in the here and now, in the shaping of each individual organism. There is a tendency to imagine that science boasts an answer to every question about the material world, but the truth is that many questions can be given several different *kinds* of answer. It is like asking how a car gets from London to Edinburgh. One answer might be 'You get in, switch on the engine, and drive up the M1.' That is not so much an explanation as a narrative, and Darwinian evolution is a bit like that—it provides a narrative that rationalizes how we got to where we are today. (As a result, it says rather little about where we might go in the future.) A chemical engineer might offer a different scenario: the car got to Edinburgh because the chemical energy of the petrol was converted to kinetic energy of the vehicle, along with a fair amount of heat and acoustic energy. That is a correct answer too, but is perhaps a little vague and abstract for some tastes. Why did the car's wheels go round? To answer this we must go instead to the mechanical engineer, who explains how they are connected to the engine via a crankshaft . . . and before long you are getting into an account of the mechanics of the internal combustion engine.

So what, we might ask, are the mechanics that create biological form? The standard answer, from what D'Arcy Thompson might have called the adaptionists, is that we must reason *a posteriori*, seeking to understand what we observe in terms of its functional efficiency and thus its adaptive value. He had no objection to this in principle, but complained that the Darwinists don't follow through with that conviction: they are happy to *assume* that a particular form must be the most efficient one, and do not bother (or do not know how) to *demonstrate* that this is so. For that reason, Thompson said, the Darwinian morphologist is apt to forget about the mechanics altogether, and to become obsessed with comparing this or that feature of different organisms without thinking how their shape is determined by their mechanical function in the organism as a whole. He argued that it was absurd to think, as comparative anatomists did, that individual bones are separately moulded by evolutionary forces, when what matters is how the skeleton as a whole functions as an efficient mechanical structure—something that can be understood using the same principles as those used by engineers to design bridges.

But this was simply an accusation that Darwinian morphologists were apt to lose the plot. The bulk of *On Growth and Form* was devoted to a much more fundamental challenge to the Darwinists' 'principle of heredity'. Thompson's contemporary zoologists would not have objected (at least, one hopes not) to the idea of invoking physics and mechanics to explain biological form, using them to rationalize it in terms of efficiency and thus adaptive advantage. But this assumes that there is a shape or structure that can be identified as the 'best' within a particular environment, and that biology can always find a way of making it. Taken to the naive extreme, this view holds that everything is possible: that nature has at its disposal an infinite palette, and that it dabbles at random with the options, occasionally (oh, so rarely!) hitting on a winning formula with which it then tinkers to make minor variations on a theme. The torpedo-and-fins theme works for fish, say, and the four-legs-and-muscle design is just the ticket for land predators.

Given the diversity of living forms evident today (Fig. 1.2), which is after all only a fraction of that which has been exhibited over geological time, this assumption of an infinite palette is understandable. But as D'Arcy Thompson reminded his readers, 'Cell and tissue, shell and bone, leaf and flower, are so many portions of matter, and it is in obedience to the laws of physics that their particles have been moved, moulded and conformed.' Evolution operates within physical constraints which insist that not everything really is possible. Are the zebra's stripes really the 'best' form of camouflage, or just the best that nature can come up with given the limitations imposed by physical law?

That might seem a minor quibble, since if the whole of nature operates within the same constraints then all we're apparently saying is that Darwinian evolution is a matter of finding the most advantageous forms out of those available. But by insisting on those limits, D'Arcy Thompson brought to the fore the issue of exactly *how* such forms come about through the action of physical forces. It wasn't just a question of ensuring that evolutionary biology obeys physical and chemical laws; he felt that these laws play a *direct, causative* role in determining shape and form in biology. Thus he insisted that there were many forms in the natural world that one could, and indeed should, explain not by arguing that evolution has shaped the material that way, but as a direct consequence of the conditions of growth or the forces in the environment.

What, he felt, could be more unnecessary than invoking millions of years of selective fine-tuning to explain the curving shape of a horn or shell when one could invoke a mathematically simple growth law, based on *proximate, physical* causes, to account for it? The sabre-like sweep of an ibex horn need not have been selected from a presumed gallery of bizarre and ornate alternative horn shapes. We can assume merely that the horn grows at a progressively slower rate from one side of the circumference to the other, whereupon, hey presto, you have an arc. In this case, then, an evolutionary argument is redundant, or at best ancillary, because the forms of horns are *inevitable*. Either a horn grows at the same rate all around its circumference, in which case it becomes a straight cone, or there is an imbalance from one side to the other, giving a curved cone. It just did not make sense to invoke other shapes: nature's palette contains just these two. Even the more elaborate spiral form of a ram's horn (Fig. 1.7) comes simply from ramping up the asymmetry of growth rates, causing the horn's tip to swing through several complete revolutions. By the same token, biological forms such as the shapes of amoeba

FIG. 1.7 Many animal horns, such as that of this Dall's sheep, have spiral shapes that can be explained with a simple growth law. (Photo: Brian Uhreen.)

can no more be regarded as 'selected for' than can the spherical form of a water droplet; rather, they are dictated by physical and chemical forces.

With Darwinism in its first flush, banishing to the dustbin of pseudo-science the teleology of Paley, it is easy to see how Thompson's ideas, with their apparent exhortation of biological predestination governed by physical law, looked dangerously close to heresy. Thompson was conscious of that, admitting that *On Growth and Form* sometimes 'undoubtedly runs counter to conventional Darwinism.' Where that was so, he said, 'I do not rub this in, but leave the reader to draw the obvious morals for himself.'

They did, and often unfavourably. At the University College of Dundee in Scotland, Thompson felt marginalized and neglected, and found himself in his sixth decade without having published anything of note. When he dared suggest, as he did in 1894 at the meeting of the British Association for the Advancement of Science, that there were 'some difficulties with Darwinism', he was met with the response that, as he put it, 'there were *no* difficulties in Darwinism... to any sensible man in those days.' Nonetheless, in the 1910s he began to draw his ideas together in what he intended to be a 'little book'. When it appeared in 1917 it was anything but that; yet the formidable *On Growth and Form* met with a largely positive reception, and it secured Thompson's reputation at last. But plans for a revised edition in the 1920s descended into farce as Thompson failed repeatedly to meet his publishers' deadlines, and the second edition, an immense tome that had to be split in two, did not appear until 1942. In some ways that delay served him poorly, for the book remained an emphatically Victorian piece of scholarship. It is illuminating to contrast it with the physicist Erwin Schrödinger's groundbreaking book *What is Life?*, published in 1944—an altogether more forward-looking work that some regard as presaging the science of molecular genetics and the notion of biology as an information science. Thompson's wartime edition, in contrast, gave the impression that little had changed in the three decades separating it from its predecessor.

And it is true that some of *On Growth and Form* has aged poorly, while some is plainly wrong. All the same, the book's central message remains relevant, and in its breadth, audacity and ambition it continues to inspire in generations of scientists (and others too) a sense of awe and wonder about the natural world. Not the least of its virtues is that it is so exquisitely written; the English biologist Peter Medawar has called it 'beyond comparison the finest work of literature in all the annals of science that have been recorded in the English tongue'. *On Growth and Form* provided the first glimpse of the landscape that we shall explore in this series. Today we can see the geography considerably more clearly, for we have the tools to map it and to interpret what we see. But we are latecomers; D'Arcy Thompson was the pioneer.

## THE GENETIC BLACK BOX

D'Arcy Thompson's thesis, then, was that biology cannot afford to neglect physics, and in particular that branch of it that deals with the mechanics of matter. (He was far less concerned with chemistry, the other pillar of the physical sciences, seemingly because he did not consider it to be sufficiently mathematical. Today there is much in the field of chemistry that would have served him well, as we shall see.) His complaint was against the dogma of selective forces as the omnipotent answer to biological questions. For him, this did not address the question of cause, but merely relocated it. A physicist, on the other hand, 'finds "causes" in what he has learned to recognize as fundamental properties . . . or unchanging laws, of matter and of energy'.

Thompson seemed to consider that this neglect of physics in biology produced an almost vitalistic streak in his contemporary Darwinists. 'The zoologist or morphologist', he wrote,

> has been slow . . . to invoke the aid of the physical or mathematical sciences; and the reasons for this difference lie deep, and are partly rooted in old tradition and partly in the diverse minds and temperaments of men. To treat the living body as a mechanism was repugnant, and seemed even ludicrous, to Pascal; and Goethe, lover of nature as he was, ruled mathematics out of place in natural history. Even now the zoologist has scarce began to dream of defining in mathematical language even the simplest organic forms. When he meets with a simple geometrical construction, for instance in the honeycomb, he would fain refer it to psychical instinct, or to skill and ingenuity, rather than to the operation of physical forces or math-

ematical laws; when he sees in snail, or nautilus, or tiny foraminif-
eral or radiolarian shell a close approach to sphere or spiral, he is
prone of old habit to believe that after all it is something more than a
spiral or a sphere, and that in this 'something more' there lies what
neither mathematics nor physics can explain. In short, he is deeply
reluctant to compare the living with the dead, or to explain by
geometry or by mechanics the things which have their part in the
mystery of life. Moreover he is little inclined to feel the need of such
explanations, or of such extension of his field of thought.

Strange that this was written nearly a century ago, for I have a feeling that
I've met this 'morphologist' and spent some time in frustrated debate with
him. He has a point: life's mechanisms are seldom simple, and (contrary to
common belief) biology does *not* always do things by the most economical
means imaginable, but can become encumbered with the legacy of history, on which physics must remain silent. Moreover, physicists can be as
arrogant as biologists are stubborn. But a tradition that has rendered
'theoretical biology' almost a term of abuse does not seem likely to be
very productive of theories that explain (rather than describe) its subject.

Of course, today's biologists have a more sophisticated answer to
questions of form than the magic word 'adaptation'. They call on genetics,
the 'microscopic' basis of Darwinism. It is tempting to imagine, from what
has been said and written about genes, that they are where all biological
questions end. One hears about genes responsible for this or that illness or
trait or feature—for cancer, for intelligence, for the development of flies'
wings or of blue eyes. The climate of the culture in molecular biology (if
not the expressed belief of all its practitioners) is that, by understanding the
roles of genes and the mutual interactions of the protein molecules they
encode, we will understand life.

That attitude underpinned the Human Genome Project, the international
effort to map out every one of the 30,000 or so genes in the 23 chromosome
pairs of the human cell. The first draft of this map was completed in 2000
(most of the holes have been filled in subsequently), and to judge from some
of the hyperbole it elicited, you would think that it has provided us with a
complete instruction manual for the human body. But it does not do that at
all. The Human Genome Project has created a bank of genetic data that is

sure to be of immense medical value, and which contains a great deal of information about how our cells work. But for biological questions that have a genetic component (and not all of them do), the respective genes are just the beginning of an answer. Most of these genes encode the chemical structures, and thus the chemical functions, of proteins. The issue is how the production (or absence) of a particular protein affects the network of biochemical processes in the cell, and how this gives rise to the particular physiological consequences that we are studying. Identifying the gene 'responsible' for this or that trait is like discovering which door we need to go through in order to enter this network and find out where it leads. (And most answers can be accessed through several doors.)

Elucidating precisely how genes work is the really hard part of the matter, which is why so much of genetics operates at the 'black box' level: we know that the presence or lack of a gene in the genome is linked to a certain manifestation at the level of the whole organism, but we do not know why. In the same way, our computers are blank boxes (available in a range of colours more tasteful than black) that we know will respond in certain ways when poked, even though most of us have not the faintest idea of why this happens.

But in any event, organisms are *not* just genes and the proteins made from them. There is all kinds of other stuff in the cell: sugars, fatty acids, hormones, small inorganic molecules like oxygen and nitric oxide, salts, and minerals such as those in bone and tooth. None of these substances is encoded in genes (that is, in the structure of our DNA), and you would never guess, by looking at the genome, that they were required at all, let alone what roles they play. And yet these substances tend to be highly organized and orchestrated in their interactions and their structures at the level of the cell (and at larger scales too). Where does that structure come from? Proteins often play a role in building it, but so too do physical forces, such as surface tension, electrical attraction, fluid viscosity. Gene-hunting tells us nothing about such things.

In short, questions in biology of a 'how?' nature need more than genetics—and frequently more than a reductionist approach. If nature is at all economical (and there is good reason to suppose that is often the case, though not invariably so), we can expect that she will choose to create at least some complex forms not by laborious piece-by-piece construction but by harnessing some of the organizational and pattern-forming phenomena we see in the non-living world. Evolution, via genetics, can exploit, tame and tune such phenomena; but it does not necessarily generate them. If this is so, we can expect to see similarities in the forms and patterns of living and purely inorganic systems, and to be able to explain them both in the same manner.

I should add a cautionary note. It is true that biology has been somewhat resistant to the idea that crude, general physical principles might sometimes be capable on their own of explaining aspects of biological form. To biologists, this seems too risky and uncontrollable, like driving a car with no hands on the wheel and hoping that friction and air resistance will somehow conspire to guide the vehicle along the winding road. But one can go too far to the other extreme. It is popular in some circles to denounce the so-called reductionist science of molecular biology and imply instead that the universe is somehow imbued with a creative potential that operates in a 'holistic' way. The fad for the notion of 'complexity', which shows that sophisticated forms and patterns may emerge spontaneously from a miasma of interactions, may sometimes veer towards a kind of neo-vitalism in the way that it invokes a cosmic creativity. Worst of all is the tendency to make moral distinctions, so that 'holistic science' becomes good and 'reductionist science' meretricious. While I applaud a perspective that broadens the horizons of 'black-box' biology and argues for a role of spontaneous pattern formation in the living world, there is no getting away from the fact that most of biology, particularly at the molecular level, is hideously *complicated*. In distinction from *complex*, this means that the details really do matter: leave out one part of the chain of events, and the whole thing grinds to a halt. In such cases, one gains rather than loses understanding by turning up the magnifying power of the microscope. Until we get reductionistic about, say, the body's immune response, we won't know much about it, let alone develop the potential to tackle pathological dysfunctions such as AIDS. A reductionist view won't necessarily provide an explanation of how it works, but without it we might not know quite what needs to be explained. Reductionism can be aesthetically unattractive, I know, but it is wonderfully useful.

## WHAT IS PATTERN AND WHAT IS FORM?

These concepts are, after all, my topics, and yet I am afraid that I cannot offer a rigorous definition of either, nor make a rigid distinction between them. If it makes you feel any better, remember that neither can scientists offer a rigorous definition of life. (They have tried often enough, but the very attempt is ill-advised, for the word is colloquial rather than scientific. You might as well try to define the word 'love'.)

What is clear is that 'pattern' is a supremely plastic word, and evidently it implies quite different things to, say, a psychologist and to a wallpaper designer. My definition, such as it is, is much closer to the latter than the former. The best I can do is to say that a pattern is a form in which particular features recur recognizably and regularly, if not identically or symmetrically. And while I shall occasionally mention patterns of a temporal nature—events that repeat more or less regularly, such as the beating of a heart—on the whole I shall be talking in spatial terms, and so the image of a pattern on wallpaper or a carpet is a useful one to bear in mind. In those cases, however, the repeating units are generally identical. My concept of pattern will not necessarily be so demanding. The repeating elements may be similar but not identical, and they will repeat in a way that could be called regular without following a perfect symmetry. Yes, I know it's vague—but I believe we usually know such patterns when we see them. One such is made up from the ripples of sand on a wave-lapped beach or in a windswept desert (Fig. 1.8). No two of these ripples are identical, and they are not positioned at exactly repeating intervals. But we can see at once that there are elementary units (ripples) that recur throughout space. We see the pattern. Indeed, we are remarkably *good* at seeing the pattern, which, because it is not mathematically perfect, is typically harder for a machine to recognize. It has become a cliché to say that the human brain is a pattern-recognizing instrument while the electronic computer is a data-crunching instrument; but like most clichés, it has taken hold for good reason. Sand ripples are a relatively simple pattern, but I think we can discern something pattern-like too even in rather irregular structures, such as the peaks and valleys of a mountain range or the skeleton of a tree in winter.

FIG. I.8   Ripples in sand clearly constitute a repetitive pattern even though no two parts of the pattern elements are identical. (Photo: Nick Lancaster, Desert Research Institute, Nevada.)

Patterns, then, are created from groups of features. Form is a more individual affair. I would define it loosely as the characteristic shape of a class of objects. Just as our brains allow us to organize a field of similar shapes into a pattern, so they are adept at somehow discerning commonalities of form between diverse objects—although we find it equally hard to explain exactly why. Objects with the same form need not be identical, or even similar in size; they simply have to share certain features that we can recognize as typical, even stereotypical. The shells of sea creatures are like this: those of organisms of the same species tend to be identifiably akin even to the untrained eye. The same is true of flowers and of the shapes of mineral crystals. You might say that the 'form' of these objects is a rather Platonic concept—that which remains after we have averaged away all the slight variations between individuals.

Patterns, then are typically extended in space—they go on and on—while forms are bounded and finite. But take this as a guideline, not a rule.

Despite our talent for spotting generic or familial similarities of pattern and form, it is not always easy to be sure that two things that look alike really do belong to the same class. To some extent, making those assessments can never be a precise science, simply because the issue may

FIG. 1.9   A checkerboard represents an orderly array of elements (*a*), while a pile of coins is disordered (*b*). (Photo: *b*, Ejdzej and Iric Zakwitnij.)

depend on what we're looking for. It is reasonable to say that a human and a chimpanzee share the same form when we are contrasting them with that of an octopus, but we can also identify chimp characteristics that differ systematically from those of humans—the ratio of arm length to leg length, say. Other comparisons are even less tangible. How can we meaningfully assess the similarities between two amorphous objects such as clouds, for instance? Yet we shall see that some apparently 'shapeless' patterns and forms have mathematically precise characteristics that do allow us to make an objective judgement about their potential kinship. Such tools are sometimes indispensable if we are looking for scientific criteria to compare different structures or to evaluate the predictions of theories of pattern formation.

I will often talk about patterns and forms being 'ordered' and 'regular', or 'disordered' and 'amorphous'. These, too, are generally rather qualitative terms, without precise definitions. When I say that a checkerboard is a very ordered pattern while a handful of coins thrown onto a tabletop is disordered (Fig. 1.9), I imagine you will know what I mean. But how might I sharpen that statement? One obvious way is to think about symmetry. A square grid is symmetrical in various ways. Technically, it possesses *symmetry operations*, which are manipulations that leave it looking unchanged. That is the case, for example, if you rotate it by 90 degrees in any direction, and also if you move the whole grid up, down or sideways by a distance equal to the width of a whole number of square cells (provided

FIG. 1.10  A mirror reflection of a checkerboard reproduces the original pattern exactly, in such a way that the two can be superimposed.

that we don't worry about the edges, or about the colour of checkerboard squares). You can also stick a mirror along various directions so that the reflection looks just like the original board (Fig. 1.10). These symmetry operations are called rotations, translations and reflections; there are other kinds, too.

You can probably appreciate that symmetry is thus related to order. But the relationship between the two is not simple. Our intuition might suggest to us, for example, that the shape shown in Fig. 1.11a is more symmetrical than that in Fig. 1.11b. But, mathematically, they both have precisely the same degree of symmetry. And yet it might be meaningful to suggest that 1.11a is more *ordered* than 1.11b, even if it is hard to find a way to define that mathematically. *Order* implies perhaps a certain logic to the construction. Formally speaking, an oak tree has no symmetry: there is no symmetry operation, other than simply doing nothing (the so-called identity operation), that will leave the shape unchanged in the sense of being superimposed perfectly on the original. But is an oak tree wholly disorganized and disorderly? I would think not. The logic of the structure, you might say, is that as you pass upwards from the trunk, it branches repeatedly at a roughly constant angle (and the limb's width decreases at each branch point).

FIG. 1.11  Which of these two shapes is the most symmetrical? In mathematical terms, they both have precisely the same degree of symmetry.

The very fact that we confuse 'symmetrical' with 'highly ordered' indicates the limited utility of symmetry as a measure of order. Is either of the shapes in Fig. 1.11 more symmetrical than a circle? Most people would say they are (this is a popular audience test in lectures on symmetry), but in fact the circle has the highest possible degree of symmetry for a two-dimensional (flat) object. There are an infinite number of rotation angles and reflection planes that leave it looking unchanged. A very high degree of symmetry can thus seem to us to be featureless and bland—quite the opposite of what we might expect intuitively.

Think, for example, of a soap bubble: to all intents and purposes it is a perfectly spherical film of liquid. Like a circle, the bubble is highly symmetrical: you can turn it this way and that, and it still looks the same. And yet now think about zooming in on the bubble until we can make out individual atoms and molecules, both in the liquid wall and the gas it encloses. Now there doesn't seem to be any symmetry at all: everything is in random disorder as molecules whiz here and there. The uniformity and high symmetry becomes apparent only by considering the *average* features of these systems. With randomness and uniformity alike, one part of the system is equivalent to any other, and things look the same (on average) in every direction. The perfect spherical symmetry of the bubble is a consequence of the average uniformity of the gas inside it, which means that the pressure it exerts on the bubble wall is equal in all directions.

The problem of creating patterns and forms that we tend to recognize as such is therefore not one of how to generate the symmetry that they often possess but of how to *reduce* the perfect symmetry that total randomness engenders (when considered on average), to give rise to the lower symmetry of the pattern. How do the water molecules moving at random in the atmosphere coalesce into a six-petalled snowflake? Patterns like this are the result of *symmetry breaking*.

The symmetry of a uniform gas can be broken by applying a force that changes the disposition of its molecules. Gravity will do that: in a gravitational field the gas is denser where the field is stronger (closer to the ground). Thus the Earth's atmosphere has a density that increases steadily towards ground level. The gas is then no longer uniform, and you can gauge your altitude by measuring the air density. Here the symmetry of the force dictates the symmetry of the distribution of matter that it produces: gravity acts downwards, and it is only in the downward direction that symmetry is broken. Within horizontal planes above the ground (more properly, within concentric spherical shells around the Earth), the atmosphere has a constant density. (Rather, it would have if the Earth were a perfect sphere and there were no winds or weather.) We might intuitively expect that this will always be so: that the final symmetry of a system will be dictated by that of the symmetry-breaking force that destroys an initially uniform state. In other words, we might expect that matter will rearrange itself only in the direction in which it is pushed or pulled, so that a pattern mimics the 'shape' of the force that generates it. Within this picture, if you want to pile up sand into mounds arranged in a square, checkerboard array, you will have to apply a force with this 'square' symmetry.

But the central surprise of the science of pattern formation is that this is not always so: the symmetry of a pattern formed by a symmetry-breaking force does not always reflect the symmetry of that force. Of the many examples that I shall describe throughout this book, one will serve here to illustrate what I mean, and why this seems at first sight to be astonishing. If you heat (very carefully—it is not an easy experiment in practice) a shallow pan of oil, it will develop roughly hexagonal circulation cells once the rate of heating exceeds a certain threshold (Fig. 1.12). The oil is initially uniform, and the symmetry-breaking force (the temperature difference between the top and bottom of the oil) does not vary horizontally. So there seems to be nothing that could make one bit of the oil behave differently from a bit slightly to its right or left. Yet suddenly this uniformity is lost, being replaced by a pattern with hexagonal symmetry. Where has this sixfold pattern come from?

Here one is apparently getting 'order for free'—getting order out without putting order in—although, as I say, it is more correct to say that symmetry is being lost rather than order gained. How is it that symmetry can be *spontaneously* broken? How can the symmetry of the effect differ from that of the cause? And why is symmetry so often broken in similar ways in apparently very different systems? That is to say, why

FIG. 1.12 When a liquid is heated uniformly from below, like a pan of water placed on the stove, it will spontaneously develop a pattern of circulating convection cells. In a well-controlled situation the cells are regular hexagons, made visible here by metal flakes suspended in the liquid. (Photo: Manuel Velarde, Universidad Complutense, Madrid.)

are some patterns universal? These are the central questions of pattern formation, and are profound enough to last us throughout the three volumes of this series.

## WHY USE MATHS?

I do not plan to say any more than this about symmetry per se, because there are many splendid books that deal with this endlessly fascinating topic, of which Hermann Weyl's *Symmetry* is a classic and *Fearful Symmetry* by Ian Stewart and Martin Golubitsky is one of the most lucid and up to date.

Neither do I propose to say a great deal more about mathematics more generally—which, without wishing to patronize, will surely come as a relief to some readers. There is, however, no escaping the fact that mathematics is the natural language of pattern and form. That may seem disappointing to those who never quite made friends with this universal tool of science—for patterns can be things of tremendous beauty, whereas mathematics can often appear to be a cold, unromantic and, well, calculated practice. D'Arcy Thompson admitted that 'The introduction of mathematical concepts into natural science has seemed to many men no mere stumbling-block, but a very parting of the ways.' But mathematics has its beauty, too, and part of that lies in the way it distils from the apparently complex an essence of pellucid simplicity. That is why maths enables us to get to the heart of pattern and form—to describe it at the most fundamental level, to reveal its Platonic core. This is more than a mere convenience; it may show us what truly needs explaining, rather than being distracted by the ephemeral or incidental. To explain how the form of a shell arises, there is no point in trying to account for every tiny bump and groove, since these will probably be different from one shell to the next. We should instead focus on the mathematical form of the 'ideal' shell.

Mathematics and geometry can describe what everyday words cannot. What is the shape of a circle? If we want to avoid tautology, we would seem to be stumped: it's no good saying it is 'round all over', since the same is true of an egg. But, geometrically, we can say that it is 'a line in a flat plane that is everywhere an equal distance from a single point'. Not

FIG. 1.13   What shape is a pebble?

only does this help us to express exactly and without ambiguity what a circle is, but it tells us how we might construct one. To draw a line equidistant from a fixed point, you can knock a nail into a wooden board and use it to anchor a piece of string tied at the other end around a pen. The geometrical description contains within it a prescription for 'growing' the object. That may seem obvious enough for a circle, but what about the shape of a pebble? Does it even mean anything to talk about a 'pebble shape', given that no two are alike (Fig. 1.13)? I would guess that you will have an immediate picture of what such a shape is like; and rather wonderfully, a team of physicists found in 2006 how to describe this shape mathematically.* This means that a theory of pebble formation by

---

*OK, you asked for it. According to Doug Durian and colleagues at the University of Pennsylvania, a pebble is a three-dimensional rounded object whose surface has a near gaussian distribution of curvature. This isn't as fearsome as it sounds. The curvature at any point on a surface is just what you might imagine it to be—a measure of how strongly curved it is. Technically, it is proportional to the inverse of the radius of a circle that fits the surface just at that point. A flat surface has zero curvature—the radius of the circle is infinite. If you measure the curvature at various points on a pebble surface and plot them on a histogram, the plot has approximately a bell-curve shape, called a gaussian distribution. This distribution is the crucial thing: it is the geometrical way of expressing the 'typical' shape, a kind of average if you will. No two pebbles will have the same curvatures at every point, or even at most points, on their surfaces, but they can still have the same overall distribution of curvatures.

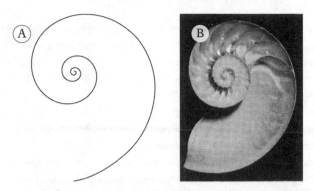

FIG. 1.14  The logarithmic spiral (*a*) is common in nature, as seen most elegantly in the cross-section of a Nautilus shell (*b*). (Photo: Scott Camazine, Pennsylvania State University.)

erosion ought to produce, from any initial rock shape, a form that converges on the Platonic pebble. The mathematical shape supplies a criterion for evaluating theories of how pebbles are made.

D'Arcy Thompson was very much preoccupied by one geometric form in particular: the so-called logarithmic spiral (Fig. 1.14*a*), which appears on a slate plaque commemorating the house in St Andrews, near Dundee, where he lived. This form was first expressed as a mathematical equation by René Descartes in 1638: the equation has a very simple and concise form, but will mean little to the non-mathematician. Crudely speaking, the spiral widens as one moves along the curve from the centre (unlike, say, the spiral made by a flat coil of rope, where the distance between each successive loop of the coil remains the same).

Thompson noted that this spiral often appears in nature, in particular in the shape of animal horns and the cross-sectional outline of mollusc shells like that of the marine Nautilus* (Fig. 1.14*b*). Its ubiquity, he said, is not at all

---

*The Nautilus has become something of the 'poster animal' of mathematical nature, but the poor creature itself gets short shrift: all we tend to see is its beautiful, empty shell. The animal is something of an oddity: a relative of squids and octopi, and highly mobile in the deep ocean by virtue of its ability to suck in water and expel it as a jet.

surprising once one recognizes that the logarithmic spiral has an important characteristic: its shape does not alter as it grows. Thus, a large Nautilus shell looks just like a small one magnified. That is just what is needed to house an organism that is steadily getting bigger in all directions. The Nautilus mollusc itself dwells in a series of chambers of increasing size, making a new one each time it has outgrown the last. It requires nothing more of each successive chamber than that it be proportionately bigger. Given that each chamber is built on the rim of the previous one, this criterion could be met by a conical shell, and indeed some molluscs do make such shells. But the Nautilus grows one part of the edge more quickly than the other, which makes the cone curl into a spiral: 'the *Nautilus* shell', said Thompson, 'is but a cone rolled up'.

Thus, the mathematical grace of the Nautilus shell does not require any geometric foresight from the mollusc, and neither does it imply that a logarithmic spiral is somehow encoded in its genes. Like the animal horns we saw earlier, the form follows in the most straightforward manner from the mode of its growth—the need to retain a constant shape which increases steadily in scale. It is fruitless, then, to argue about why the logarithmic spiral is somehow superior to others in evolutionary terms—it is simply a consequence of the mathematics of growth.

This perspective can be extended to spiral forms considerably more complex than that of the Nautilus shell. D'Arcy Thompson realized that other shells too have a shape that can be generated by the logarithmic spiralling of a particular rim shape, later called the 'generating curve'. He said:

> The surface of any shell may be generated by the revolution about a fixed axis of a closed curve, which, remaining always geometrically similar to itself, increases its dimensions continually ... The scale of the figure increases in geometric progression [that is, being multiplied by a constant factor on each growth step] while the angle of rotation increases in arithmetical [that is, at a constant rate].

The process is shown in Fig. 1.15*a*, and the forms it can produce have been explored in computer modelling by Deborah Fowler and Przemyslaw Prusinkiewicz at the University of Regina in Canada (Fig. 1.15*b*). Thompson noted that the shape of the generating curve 'is seldom open to easy

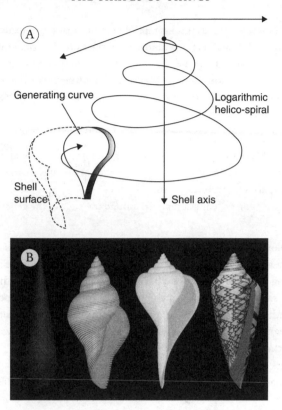

FIG. 1.15 A shell surface can be constructed by sweeping a two-dimensional 'generating curve' through a logarithmic spiral (*a*). This procedure will create many different types of shell surface (*b*), depending on the generating curve. Here some of these artificial shells have been given surface pigmentation patterns for added realism. (Image b: Przemyslaw Prusinkiewicz, University of Calgary.)

mathematical expressions', but the way the shell is created by sweeping this boundary in a spiral around a fixed axis *is* mathematically well defined, and can be seen to be a consequence of a simple growth law.

Thompson admitted that in general it was no easy thing to find mathematical expressions that describe the organic forms of nature, and on the whole he was right. But as this example illustrates, that is not really the right way to proceed. Equations describing the surfaces of the shells in Fig. 1.15b would indeed be cumbersome, and probably not very illuminating. It is far more instructive to look for the *algorithm* that generates the form—for a mathematical description of how to *grow* it. An algorithm is a series of steps conducted in succession. In this case, it could read:

> Step 1: Choose your generating curve.
>
> Step 2: Move the generating curve along a logarithmic spiral, while letting it grow bigger at a steady rate and, optionally, at the same time descending a vertical axis.
>
> Step 3: As you go, deposit material around the edge of the generating curve.

Complex shapes and patterns of the sort I consider in this book are often most easily described not in terms of 'what goes where' but by a generating algorithm. Once we have identified an algorithm that makes the right shape, we can ask which physical processes might produce such an algorithm. If the algorithm gives the right shape, that doesn't necessarily mean it corresponds to anything that happens in the real world—but at least it *could* do.

## MODEL MAKING

When scientists construct an algorithm like this, they have an example of what they often call a 'model' of the system they are studying. They will say 'Here is my model of shell growth', and will outline the sequence of steps in the process. This is perhaps a rather unfamiliar and technical use of a word that, in everyday parlance, has a somewhat different (but related) connotation. We tend to think of a model as a small-scale replica of a real object (the definition remains apt for fashion models). For scientists, however, a model is a simplified, abstract description of what they think is going on in a particular phenomenon. The model is basically a series of assumptions about what is involved in the process, which the scientist proceeds to translate from qualitative to mathematical terms, before performing calculations to see if the model predicts an outcome anything like that actually observed.

The key here is the word 'simplified'. A model does not generally take exhaustive account of everything that might be happening in the process. A good model includes only those aspects thought to be essential in producing the basic phenomenon. The details may be added later, once the basic process is understood. There are various reasons why scientists prefer to simplify their models. Sometimes they just don't know everything that might be happening in the system being studied—this is almost invariably the case when the system is alive, for example. Or it may be that some factors are clearly going to have only a minor influence, so that their inclusion just makes the equations harder to solve without altering the solutions very much. If you want to calculate the speed of a cannon-ball falling from the Tower of Pisa, you needn't worry too much about including a mathematical description of air resistance; but if you're looking instead at small ball-bearings falling through treacle, the fluid's resistance to the motion is crucial. A lot of engineering is conducted in this spirit: keep what matters, forget the rest (at least in the first instance). There is a third reason for leaving things out of a model, however: we might know, or at least suspect, that they are probably important, but we simply do not know how to include them, or how to solve the equations if we do. So we might then look for reasonable approximations that we *do* know how to solve, accepting that what the model predicts might not match very well with what we observe. Scientists interested in fluid flow, for example (the topic of Book II), commonly find themselves in this situation, making big approximations and having to live with the consequences.

The point is that scientific descriptions of phenomena do not fully capture reality, nor do they aim or claim to. They are models. This is not a shortcoming of science but a strength, since it allows scientists to make useful predictions without getting bogged down by intractable details. Much of the scientist's art lies in working out what to include and what to exclude in a model.

There are very many natural phenomena (one could make a case for this being true of them all) for which there is not a single, unique model that is 'right'. This is more than a matter of models differing by the choice of what to put in and leave out. Rather, some phenomena can be tackled successfully from more than one entirely different theoretical perspective. For example, one can write down equations that describe traffic flow on a road rather as though it was a liquid flowing down a pipe. Or you can write a computer programme that takes account of the behaviour of each individual and discrete vehicle, acting out the flow rather like a computer game. Both models can show some predictive value—both are valid models. This is true of several of the phenomena that I shall discuss in this book, and it means that it can be rather hard to decide the relative merits of a particular model. If two seemingly different models both capture aspects of the real system under study, which is best? There may be no right answer—perhaps one model is best for one purpose, and another for understanding some other aspect. That is a reminder that science is about finding good enough approximations to reality to give us some understanding of it, and not about capturing an absolute 'truth'.

Some models are expressed in terms of mathematical equations, for example describing the forces at play (such as Newton's inverse-square law of gravity). If the modeller is lucky, it may even be possible to solve such a model with pen and paper (which was all theorists had at their disposal until half a century ago). If the calculations are too hard, they can be made on a computer. But other systems might not be so amenable: you know how the components interact, but not how to describe that interaction with equations you can solve. In such cases, one can conduct a simulation of the system on a computer, such as that described above for traffic. The predicted behaviour emerges not by solving any maths but by running the simulation and seeing what happens.

Perhaps the strongest point that I want to make about models in the present context is that they can often generate the complex patterns seen in nature from remarkably few ingredients, which are themselves of striking simplicity. What does that tell us? On one level it simply means that growth and form need not be mysterious—we do not have to resign ourselves to thinking that the shape of a flower will be for ever beyond our abilities to explain, or even that an explanation (at some level) will require years of dedicated research on plant genetics. On the other hand, it carries at least an implication that there exist universal patterns and forms that remain insensitive to the fine details of a particular system. Bear this idea in mind as we peruse the gallery of extraordinary and often beautiful patterns and forms in these three books: we will see that nature, like any artist, has themes and preferences for the images it creates. For nature, at least, we can sometimes understand why this is.

## THE MAP IS NOT THE TERRITORY

Models are maps of reality: they include only the features one wants to study, and leave out everything else. Maps have a fascination of their own, but they are as nothing compared to the real thing—to a walk in the woods and mountains. That is why I recommend that you try to create some of these patterns for yourself, with the recipes given in the appendices. I hope you will discover that the most exciting, the most profound experience of them is to be found through direct encounter. This is not so hard to arrange, for these self-made patterns are everywhere, in the vegetable patch, in the coffee cup, on mountain tops, and in the city streets. I hope you enjoy them.

# LESSONS OF THE BEEHIVE

## Building with Bubbles

A photograph of Ernst Haeckel doing field research on Lanzarote in 1866 shows the archetypal German Romantic: here are the abundant waves of glossy hair, the full beard, and the distant gaze of the dreamer. This image is furthered in a letter that Haeckel, a professor of zoology at the University of Jena, wrote to his parents from the Spanish island early the next year, in which he describes a jellyfish called a siphonophore:

> Imagine a delicate, slender flowering whose leaves and brightly coloured flowers are transparent like glass, and which meanders through the water with the most graceful and sprightly movements, and you have an idea of these delightful, beautiful and delicate examples of animal finery.

The impact that this 'finery' had on Haeckel is revealed in dazzling abundance in the portfolio *Art Forms in Nature* which he published in ten instalments over five years, beginning in 1899. It looks at first glance like the original coffee-table book, a collection of one hundred glorious plates drawn by Haeckel himself, in which he depicts the profusion of wonderful forms found among living things. Here are antelopes and birds, turtles and crabs, lichens and pine cones—but also, and in greater numbers, organisms that most people never knew existed, and still don't. There are the delicate fronded forms of *Medusae* jellyfish—the siphonophores and their relatives (Fig. 2.1)—and all kinds of strange corals and snails and sea creatures that look like the most extravagant inventions of a Surrealist. Most strikingly, there are forms that do not look like living organisms at all: one might imagine they are elaborate, ornamented shields and plates, futuristic spacecraft covered in spines, bizarre crown-like headgear, de-mented pavilions (Fig. 2.2).

There are no clues here to the scale of these objects, nothing to show that the remarkable spiny cages and domes are in fact the 'shells' of single-celled organisms visible only under the microscope. They are called radiolarians, and Haeckel's study of these creatures under the supervision of the physiologist Johannes Müller in Berlin in the 1850s led to his 1862 *Monograph on Radiolarians*, which helped to gain him the post at Jena. 'When I visited the ocean for the first time twenty-five years ago', he wrote in 1879 in a book on jellyfish,

> and, in August 1854, was introduced to the inexhaustible, wonderful world of marine life by my unforgettable master Johannes Müller on Heligoland, nothing exerted such a powerful force of attraction on me amongst the myriad animal forms, of which I had not seen living specimens until then, as the medusae. Never will I forget the delight with which I, as a twenty-year-old student, first observed Tiara and Irene, Chrysaora and Cyanae, and attempted to render with my paintbrush their splendid forms and colours.

Marine biology, more or less unfettered by gravity, seems to luxuriate in these baroque forms, and it is not hard to see how a young man, an admirer of Goethe and enthused with the spirit of Romanticism, and a gifted artist to boot who had even considered a career as a painter, might, after gazing at them for countless hours, be moved to conclude that even the most basic living matter has an 'artistic soul'. We can understand why he might feel compelled to communicate this natural artistry to a wide audience with such ornate and mesmerizing drawings. But Ernst Haeckel was not the kind of person to let it rest at that. In many ways he is the closest thing to a German D'Arcy Thompson—a synthetist, a man who sought to find common threads running through the profusion of nature's tapestry. In other ways, however, he could not be more different.

37

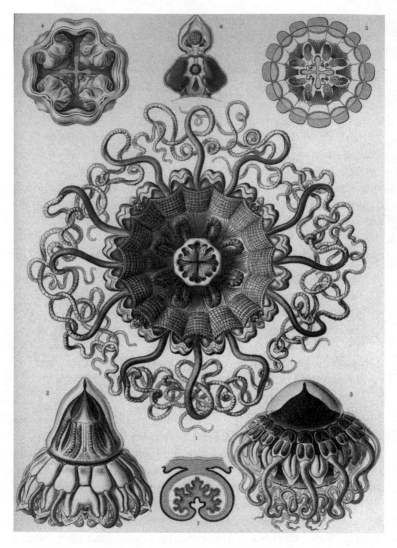

FIG. 2.1 *Medusae* jellyfish drawn by Ernst Haeckel.

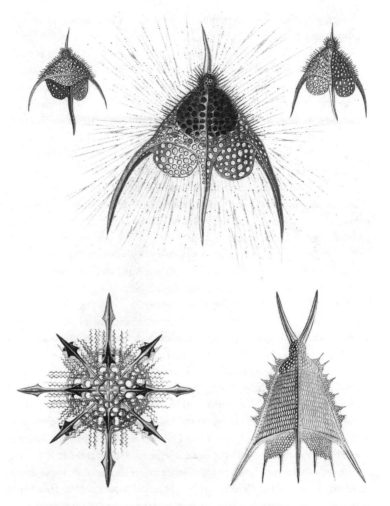

FIG. 2.2 Some of the radiolarians depicted by Haeckel look like the products of a baroque imagination.

Thompson was the embodiment of Scottish rationalism, an engineer at heart who looked for proximal, mechanical explanations of natural forms. Haeckel was close to being a mystic, finding within the traceries of a radiolarian evidence of a creative, organizing force that pervaded nature, and hanging on these tiny mineralized skeletons a thesis about aesthetics, sociology, and religion.

We need to look at *Art Forms in Nature* with this in mind. For it is not a casually assembled cabinet of curiosities, but a tract with an agenda. Haeckel has selected carefully what he shows us, and how. These are not images to massage the senses (although they certainly do that); they are, in Haeckel's mind, raw data marshalled to convince us of his overarching theory about form, symmetry and beauty in the world.

Haeckel was a Darwinist—indeed, Darwin himself credited Haeckel as being instrumental to the spread of his theory of evolution in Germany—but his creed was not exactly orthodox. He was convinced that organisms evolve by degrees and thereby diversify in a branching phylogenetic tree, achieving forms of ever-increasing complexity. But he did not feel that natural selection was necessarily the mechanism; indeed, his insistence that the environment helps to shape an organism sounds surprisingly close to Lamarckism, the notion of inheritance of acquired characteristics. And Haeckel's evolutionary theory did not trust wholly to the randomness and contingency of Darwin's idea, for he discerned something more deterministic at the root of it all. Influenced by Hegel's view of history as spiritual destiny, Haeckel implied that there was an organizing force that shaped nature all the way from protoplasm (primordial living matter) to humankind. Indeed, he felt that even our mental and spiritual states are direct products of evolution: somewhere along this great chain of being we acquired a soul. Our appreciation of beauty and our capacity for mathematical abstraction both emerged from this tendency of nature to organize itself into ever more complex forms, based on symmetry and pattern. Thus Haeckel held a somewhat Platonic view of beauty, which is not in the eye of the beholder but is the result of the way we are conditioned by the objective forms of the natural world. In his bestselling *Riddle of the Universe* (1899), Haeckel asserted that there is a fundamental unity between organic and inorganic matter, not simply because (as some chemists had argued already) they are made of the same basic stuff but because the drive towards organization, and thus towards life, is inherent in all matter. When in 1888 the Austrian botanist Friedrich Reinitzer discovered the first liquid crystal—a compound whose molecules became spontaneously aligned with one another in the liquid state, rather like logs floating on a river—Haeckel was delighted. Here, he believed, was evidence of simple matter exhibiting nature's organizing principle, and in his last and perhaps strangest book, *Crystal Souls: Studies on Inorganic Life* (1917), he argued that liquid crystals are a genuine kind of life.

This all amounts to a rather peculiar distortion of Darwin's grand idea. But *Riddle of the Universe* is disturbing not just because of its teleology. It also shows abundant evidence of the darker side of Teutonic Romanticism. The Cambridge evolutionary biologist Simon Conway Morris says of the book,

> Vastly popular, endlessly reprinted and translated, it nevertheless 'appealed to a pseudo-educated mind ... without much sophistication who had sought an authoritative yet simple account of modern science and a comprehensible view of the world'. Behind the bearded sage and devotee of the little town of Jena, was an intolerant mind wedded to racism and antisemitism ... His farrago of ideas ... found a warm reception with the Nazis. Just how much Hitler knew of Haeckel's actual work is not clear, but the influence of his philosophy is obvious.

The role of Haeckel's *faux*-Darwinian ideas, and of the so-called Monist league that he founded in 1906,* in the emergence of European fascism has been explored by the American historian Daniel Gasman. It makes the eugenic enthusiasms of the early British Darwinists seem mild in comparison, and serves as a warning about the dangers of turning science into ideology.

## FROM NATURE INTO ART

The gorgeous plates in *Art Forms in Nature* were thus intended not as mere demonstrations that 'nature is beautiful' but as *data* supporting Haeckel's contention that the emergence of spontaneous form, symmetry, and order in the living world was an inevitable process. This should make us look at the pictures with a more critical eye. When Haeckel wants to show us nature's organizing force at work, we must start to wonder if he is doing a little organizing of his own. Do the fronds and globular chambers of

---

*Monism was the idea that all life stems from matter alone, denying the dualism of body and spirit. Haeckel pursued this idea throughout his career, leading him to be regarded as anti-religious and atheistic. The link between Monism and Nazi ideology is still debated; at the very least, one must say that the equation is not simple.

medusae really have such perfect forms and symmetries, as though they are the engineered products of an architect's blueprint? His cofferfish are Platonic creatures with polygonal scales, and Haeckel has plucked off some of those scales and served them up as abstract geometrical designs. The heads of his bats have the uncanny bilateral symmetry of Rorschach blots. The fossil sea creatures called cystids look here like elaborate caskets fashioned by silversmiths, unreal in their precision. Haeckel was a wonderful draughtsman, but was he really drawing what he was seeing, or what he felt he should be seeing—the idealized form that he intuited behind the debased, mundane reality?

First and foremost, Haeckel's illustrations have a decorative quality of a now familiar stamp. The fronds of his discomedusae (Fig. 2.3) resemble nothing so much as the leaves and stems of a William Morris floral print, while the swirling exuberance of their trailers make us think of the artistic movement that was very much in sway when Haeckel was making these drawings: Art Nouveau and its German equivalent, Jugendstil. That is no coincidence, for Haeckel was both influenced by this movement and influenced it in turn. His jellyfish images were used as ceiling decorations in his house, the Villa Medusa (Fig. 2.4), where they were perfectly in keeping with the Art Nouveau furnishings. Haeckel's pictures influenced artists such as Hermann Obrist and Louis Comfort Tiffany, and their impact was acknowledged most explicitly in the work of the French architect and designer René Binet, who wrote to Haeckel in 1899 while working on the entrance gate to the Paris World Exposition of 1900:

> About six years ago I began to study the numerous volumes written about the Challenger Expedition [see below] in the library of the Paris museum and, thanks to your work, I was able to amass a considerable amount of microscopic documentation: radiolarians, bryozoans, hydroids etc. . . . , which I examined with the utmost care from an artistic standpoint: in the interest of architecture and of ornamentation. At present, I am busy realizing the monumental entrance gate for the exhibition in the year 1900 and everything about it, from the general composition to the smallest details, has been inspired by your studies.

We can see that clearly enough in the result: Binet's gate is an ingenious adaptation of Haeckel's radiolarians (Fig. 2.5). Two years later, Binet

FIG. 2.3 Haeckel's *discomedusae* seem to owe a clear debt to the arabesque Art Nouveau style.

FIG. 2.4 The dining room ceiling at Haeckel's Villa Medusa acknowledges the ornamental style of his drawings.

FIG. 2.5 René Binet's design for the entrance gate to the Paris World Exposition of 1900 was inspired by Haeckel's drawings of radiolarians.

expanded on this theme with a book of Art Nouveau designs entitled *Decorative Sketches* (Fig. 2.6), in regard of which he said to Haeckel, 'The book that I will be publishing will clearly demonstrate the high value of your works, and it will assist those, who do not know very much about the history of these infinitely small creatures, to understand the significance of "artistic forms"'.

To Binet it seems that this 'significance' lay with the inspiration that Haeckel's arabesques could offer the artist. But Haeckel was more concerned with arranging those shapes into a theory of natural form. He was the original 'morphologist', a scientist who sought order among the myriad forms that nature presents. Central to his theory was the notion of a systematic progression in the structures that matter adopts, from the simple to the complex, always guided by principles of geometry that echoed those evident in crystals. Darwinism fitted this theory of morphology (or we might say rather that Haeckel made it fit) because it implied precisely this kind of genealogical 'unfolding' through which nature's organizing principle was increasingly elaborated. Haeckel supposed that the rules of this unfolding were universal, and this led him to the startling idea that the same forces that guided protoplasm through successive evolutionary stages over geological eons were also responsible for organizing a single-celled egg into a complex organism during its growth, a process called ontogeny. In Haeckel's epigrammatic phrase, 'ontogeny recapitulates phylogeny'. In this view, the developmental stages that a human embryo passes through represent a kind of speeded-up replay of evolution, so that, for example, the fetus exhibits a gilled, fish-like state on its way to becoming a human. Thus, Haeckel argued, the primitive embryonic stages of all complex organisms are essentially identical—a proposal that he supported with drawings based on the evidence available at that time. Again, Haeckel seems to have carefully embellished and adapted this visual evidence to suit his case, which has earned him much opprobrium in later times as an apparent fraudster who fixed his data.*

But whether or not Haeckel was sometimes liberal with the evidence, his 'biogenetic law' of phylogeny and ontogeny was prescient in that it repre-

*The truth of that is complicated: see, for example, Nick Hopwood's analysis in *Isis* 97 (2006): 260–301.

FIG. 2.6 The designs in Binet's book *Decorative Sketches* also draw on Haeckel's visual vocabulary.

46

sents one of the earliest searches for general organizational principles that imprint pattern on the growth and form of organisms. This is the topic to which I shall return at the end of the book. I want now to turn to Haeckel's early passion: radiolarians. In the geometric skeletal shells of these single-celled sea creatures there seemed to be the most explicit demonstration of a mathematical ordering principle in operation. In fact, Haeckel sought a classification of radiolaria shells based on the kinds of symmetry schemes used to describe crystal forms, and a page from his *General Morphology of Organisms* makes clear just how Platonic and geometrical his thinking was here (Fig. 2.7). This 'organic crystallography' was the foundation of his attempt to develop a classification of all living things using the geometrical properties of their forms, a scheme that he called organic stereometry.

Yet Haeckel, the Hegelian prophet of a grand unified theory of life based on a kind of 'will to order', was less concerned about what was, to D'Arcy Wentworth Thompson, the most pressing question of all: by what mechanical process do structures as intricately patterned as this become formed?

## HONEYCOMBS OF THE SEA

When the complex shells of microscopic marine organisms were first discovered, it was hard to believe that these regular, geometric forms were a product of the organic world at all. Radiolarians are not the only such organisms with delicately patterned shells (more properly called external skeletons or exoskeletons). There are diatoms, silicoflagellates, dinoflagellates, coccolithophores, and others (Fig. 2.8). In 1703 an anonymous contribution to the *Philosophical Transactions of the Royal Society* describes 'pretty branches, compos'd of rectangular oblongs and exact squares' that adhere to the roots of pond weed, visible only under the microscope: these are species of diatom, many more of which were identified in the late eighteenth century. But no one was sure whether they should be classed as animals or plants. In fact, diatoms, like dinoflagellates and coccolithophores, are types of phytoplankton: microscopic aquatic plants. Radiolarians, on the other hand, are primitive, single-called animals called protozoans. All of them build mineral exoskeletons for defence, which is why they are often elaborated with spines and spikes like tiny marine hedgehogs. Radiolarians, diatoms, dinoflagellates, and

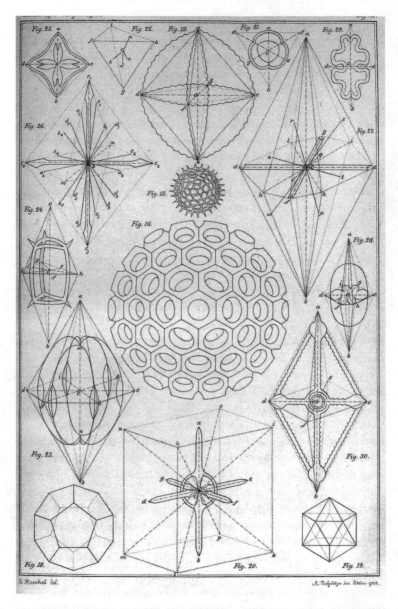

**FIG. 2.7** Haeckel sought to classify radiolarian geometry using the symmetry principles that were applied to crystal shapes.

48

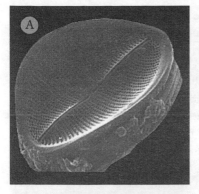

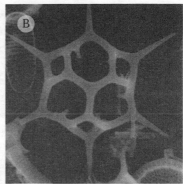

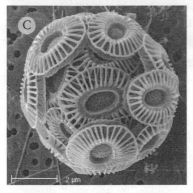

FIG. 2.8 Ornate exoskeletons are common in aquatic micro-organisms, such as diatoms (a), silicoflagellates (b), and coccolithophores (c). (Photos: a, Rex L. Lowe, University of Hawaii; b, Kevin McCartney, University of Maine at Presque Isle; c, Jeremy Young, Natural History Museum, London.)

silicoflagellates make these deterrent shells from silica (silicon dioxide, the fabric of quartz and window glass), whereas coccolithophores, which are abundant in warm, tropical seas, fashion their patterned plates and shells from calcium carbonate, the stuff of chalk and marble. Indeed, both chalk and marble are *derived* from the shells of such micro-organisms, deposited as sediment on the sea bed when the organisms die: the soft tissues decay, but the 'bones' remain. Many of these shells can still be discerned in pieces of chalk inspected under a microscope.

That was how the German biologist Christian Gottfried Ehrenberg studied coccolithophores in the 1830s. When he first saw them in chalk from an island in the Baltic Sea, Ehrenberg could not believe that these oval plates with their regular radial patterns (Fig. 2.8c) were the products

49

of living organisms, and he imagined that they must be inorganic formations, examples of crystals known as spherulites. He called them 'calcareous morpholiths', and like Haeckel he recorded what he observed in
meticulous drawings. These appear among the five thousand or so illustrations in the first volume of Ehrenberg's great work *Microgeology* (1854),
where they are still presented as inorganic crystals. Ehrenberg also did
most of the important early work on radiolarians and flagellates, inspect-
ing their fossil forms in sedimentary rock.

Thomas Huxley, Darwin's indefatigable supporter, came to the same
conclusion about coccolithophores when in 1857 he saw 'rounded bodies'
in sediments dredged from the North Atlantic. These, said Huxley, looked
'somewhat like single cells of the plant *Protococcus*', a kind of algae; but,
nonetheless, Huxley presumed that they were mineral structures, and he
proposed that they be called coccoliths: 'coccus stones'. Others were not
so sure about that. Ehrenberg and Huxley saw only the isolated platelets
that overlap to make up the shell. But in 1860 the naturalist George
Wallich investigated Atlantic mud collected by the British ship HMS
*Bulldog* and saw within it coccoliths assembled into balls, which he called
coccospheres. He supposed that these shells encased the larval forms of
marine plankton, called foraminifera. Meanwhile, the British geologist
Henry Clifton Sorby also found fully assembled coccospheres in English
chalk, and he noted that the platelets were not flat but convex on one side
and concave on the other. That did not look like the result of simple
crystallization, but suggested that the coccoliths were shaped around a
spherical form: they were, he decided, the shells of organisms. Huxley was
persuaded of the biological origin of coccoliths, and he went on to claim
that in marine sediments they were often embedded in organic slime,
which he identified as the primal living matter called protoplasm that
Haeckel had described. In recognition of this, Huxley named this new
species *Bathybius haeckelii*. Haeckel, naturally, was delighted with the
discovery, and announced that *Bathybius* was the fundamental life form
from which all others were derived—the original Ur-matter of life, preg-
nant with shape and form. On the stable, unchanging sea bed, he said,
*Bathybius* slime had persisted essentially unchanged since the dawn of time.

But this protoplasm was not really a new species of organism at all. It
turned out that Huxley's jelly was simply the product of chemical

reactions between the sea water and the alcohol used to preserve the specimens. *Bathybius*, said Huxley lugubriously on hearing the news, 'has not fulfilled the promise of its youth.' All the same, the notion of coccoliths as the protective shell of soft organisms was correct, and in 1898 George Murray and V. H. Blackman proposed that they be named coccolithophores. Ehrenberg himself, however, resisted all this talk; to him, coccoliths were inorganic and that was that, and he defended this position staunchly until his death in 1876.

Scientists and naturalists came to appreciate the tremendous diversity of marine exoskeletons as a result of the transglobal voyage made between 1872 and 1876 by the British research vessel HMS *Challenger*, a landmark expedition for oceanography. The crew of *Challenger* measured the ocean's temperature and depth and mapped out the sea bed as the ship sailed for nearly 70,000 miles across the Atlantic and the Pacific. Another aim of the mission was to examine the organisms dwelling in the deep sea, and the dredges brought to light all manner of radiolarians, coccolithophores, and other organisms. Haeckel seized on this bounty, cataloguing the forms of radiolarian exoskeletons in a vast atlas (Fig. 2.9).*

This was rich material for D'Arcy Thompson. He found in Haeckel's catalogue of radiolarians a treasure trove, an almost overwhelming parade of nature's architectural inventiveness begging for explanation. One of the forms on which Thompson fixed his attention was in some ways the least elaborate: free of spines and shaped into an apparently perfect sphere, the shell of *Aulonia hexagona* was a structure of geometric exactitude (Fig. 2.10), looking, in Thompson's words, 'like the finest imaginable Chinese ivory ball'.

*A. hexagona* is named for its mineral cage, which appears to be composed entirely of a mesh of hexagons. But look more closely and you will see here and there some pentagons in the lattice too. These, as Thompson explained, were not the result of faulty manufacturing by the organism that created the dome, but were in fact essential in order to make a dome in the first place. For, he wrote, 'as we learn from Euler: the array of hexagons may be extended as far as you please, and over a surface either

*Ironically, it was *Challenger's* chemist, John Young Buchanan, who debunked Haeckel's *Bathybius* protoplasm.

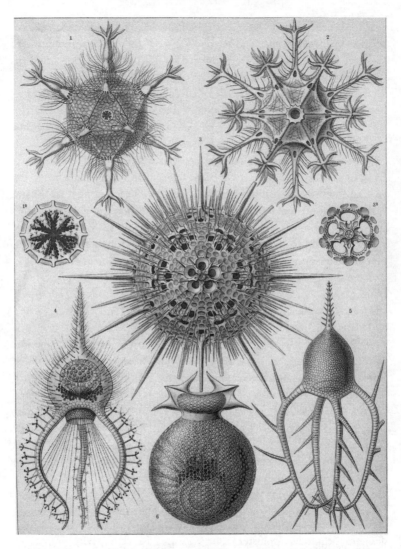

FIG. 2.9 The *Atlas* of Ernst Haeckel presents a vast array of beautiful radiolarian exoskeletons.

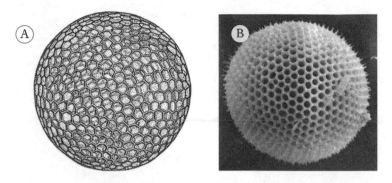

FIG. 2.10  The radiolarian *Aulonia hexagona* as drawn by Ernst Haeckel (*a*), and as it appears in the electron microscope (*b*). (Photo *b*: Tibor Tarnai, Technical University of Budapest.)

plane or curved, but it never closes in.' No network composed only of hexagons can ever form a closed shell like this. That is implied by a formula deduced in the eighteenth century by the Swiss mathematician Leonhard Euler, which describes the relationship between the number of faces, edges, and vertices of any closed-up polyhedron. Euler's formula tells us that you cannot make such a polyhedron from hexagons alone. It is no good taking a sheet of chicken wire and trying to bend it into a dome— you'll never succeed (unless you grossly distort some of the hexagonal elements). But introduce a few pentagons, and the problem starts to look soluble, because pentagons make the sheet curve (Fig. 2.11). As Thompson pointed out, Haeckel seemed to have some inkling of this: his description of *Aulonia hexagona* does mention that there are a few pentagonal, and indeed square, facets in the radiolarian's shell.

The number of pentagons needed to bend a sheet of hexagons into a closed shell is not arbitrary. Euler's formula tells us how many are needed, and the answer is precisely twelve. It does not matter how many hexagons are included; so long as you have twelve pentagons, you will 'achieve closure', as the popular expression has it. The American architect and designer Richard Buckminster Fuller knew this, which was why his famous geodesic domes of the 1950s and 1960s incorporated strategically placed pentagons among the mesh of hexagons (Fig. 2.12). But Euler's formula is not known as well as it should be in the sciences. When a team of chemists deduced in 1985 that carbon atoms, which link up into flat sheets of

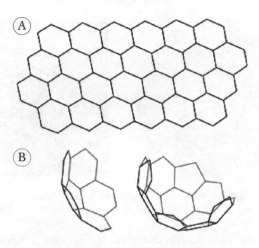

FIG. 2.11   A sheet of hexagons is flat (a), but if pentagons
are added then it is forced to curve (b).

hexagons in graphite, form stable clusters of 60 atoms each, it took hours of frustrating trial and error before one of them thought to consult Buckminster Fuller's work and thereby discovered how pentagons would allow the sheets to curl up into hollow balls. They named this carbon molecule buckminsterfullerene in the architect's honour.

For D'Arcy Thompson, unlike Haeckel, the problem presented by radiolarians could not be explained by some vague ordering principle with which all matter is imbued. Nor did it seem very satisfactory to suppose that *Aulonia hexagona* has been equipped by natural selection with a kind of hexagonal-mesh-making machinery that includes instructions to insert precisely twelve pentagons. The problem is, rather, how an individual organism, possessed of the faculty for condensing a mineral out of the dissolved ingredients in sea water, can arrange that hard material into such an organized pattern. What is the mechanical process that produces such a form? Thompson reasoned that such a structure is by no means unprecedented in nature: we can see it most clearly in the bee's hive. And there is no need to regard it as anything uniquely *biological*, for it is the easiest thing in the world to make such a pattern in non-living matter. All we need are *bubbles*.

FIG. 2.12 Richard Buckminster Fuller used hexagonal and pentagonal elements to construct his geodesic domes, such as that used in the US exhibit for Expo '67 in Montreal. (Photo: Copyright 1967 Allegra Fuller Snyder, courtesy of the Buckminster Fuller Institute, Santa Barbara.)

## WATER'S SKIN

*On Growth and Form* has many pages devoted to cellular patterns with hexagonal symmetry, in which D'Arcy Thompson asserted that 'The most famous of all hexagonal conformations, and one of the most beautiful, is the bee's cell.' The honeycomb is indeed a natural pattern of enduring fascination. Those master geometers the Egyptians kept bees 5,000 years ago, and would surely have been entranced by the polygonal array of compartments in which they store their honey, harvested pollen, and eggs (Fig. 2.13). We have no record of what they made of the bees' work; but the *Arabian Nights* tells us that Euclid himself, who allegedly taught the Egyptians the geometric arts, 'was instructed by admiring the geometry of their cells'. Around the third century BC, the Greek mathematician Pappus of Alexandria explained that the bees had only three choices for making

FIG. 2.13 The hexagonal honeycomb of the honey-bee was surely one of the first recognized examples of geometrical pattern in the natural world. (Photo: Karunaker Rayker.)

identical cells with perfect polygonal cross-sections that would pack together in space without any gaps: squares, triangles, and hexagons. Of these, said Pappus, 'the bees have wisely selected for their structure that which contains most angles, suspecting indeed that it could hold more honey than either of the other two.' Thus, Pappus concluded, the bees are possessed with 'a certain geometrical forethought'.

Pappus does not make himself terribly clear, however, for if you want a cell to hold more honey then you can simply make it bigger, whether it is hexagonal, square or indeed star-shaped. The eighteenth-century French scientist René de Réaumur made a more thorough statement of the problem, pointing out that what matters is not the volume of the cavities but the area of the walls: to fill a given area of space with hexagonal cavities requires a smaller total length of cell wall than to do so using triangular or square cells of the same cross-sectional area. In other words, making hexagonal honeycombs lets the bees economize on wax, the production of which costs them metabolic energy. This need to economize on labour costs makes perfect sense within a Darwinian world of selective pressures, and indeed Darwin declared the honeycomb 'absolutely perfect in economising labour and wax'.

But this was just the kind of Darwinian fable that left D'Arcy Thompson fuming. Let's think about what it entails. First, we must assume that the ancestors of today's honey-bees tried out just about every comb geometry possible—squares, triangles and, who knows, perhaps jigsaw-piece shapes too. Those making hexagons completed the job in short order and were soon out again foraging for pollen while the others were still struggling to cook up enough wax. But then the hexagon-makers had to pass on this geometric advantage to their progeny, presumably by means of genes that encoded complex measuring apparatus for constructing ideal hexagons without the assistance of compasses, protractors, and copies of Euclid's *Geometry*.

Why accept this concoction of untested and complicated suppositions, Thompson argued, when there was a much simpler explanation for the hexagonal cells that did not invoke Darwinism but only the laws of physics? The seventeenth-century Danish mathematician Erasmus Bartholin may have been the first to propose something of the sort. Imagine a horde of bees each constructing its own cell and striving to make it as large as possible, he said. Then each compartment presses up against those of the neighbouring bees, and what you have is akin to a raft of bubbles pushing against one another. And as everyone who has played with a straw and soapy water knows, the bubble raft is a hexagonal array (Fig. 2.14). D'Arcy Thompson suggested that the 'pressure' of each bubble did not have to come from the bee pushing on its walls; rather, if their combined body heat renders the wax soft, then the cells are indeed like bubbles with walls of sluggish liquid, which will be pulled into a hexagonal array by surface tension.

This all sounds plausible enough, perhaps, but it does not fully explain the hexagonal pattern in any fundamental way—it simply says that a honeycomb is like a bubble raft, and bubble rafts make hexagonal arrays. Why hexagons and not squares, or crazy pavings of random polygons? If we want a more complete explanation, we need to know what bubbles are and what controls their shape.

A bubble is a volume of gas surrounded by a liquid. In champagne the bubble is embedded in liquid, while soap bubbles have thin liquid walls. In both cases the bubbles are spherical and, as I suggested earlier, this can be considered a consequence of the fact that the gas inside the bubble pushes equally on the wall in all directions. But there is more to it than that. The spherical shape is robust: if deformed, the bubble becomes a sphere again very quickly. The force that pulls it back to this shape is the surface tension.

FIG. 2.14 A bubble raft of
equal-sized bubbles adopts
the hexagonal pattern of
a honeycomb. Is this a
coincidence? (Photo:
B. R. Miller.)

A liquid's surface tension arises from the simple fact that the surface is
where the liquid stops. The atoms or molecules that it consists of cohere
to one another by relatively weak electrical forces of attraction, rather like
those that pull hairs towards a comb that has passed through them. A
molecule of water deep within the liquid feels these attractive forces
coming from all directions because of the molecules that surround it.
But for molecules at the water surface, these attractions come only from
below and to the side, not from above (Fig. 2.15). So there is an overall
inward force on the surface molecules, which results in surface tension.

These attractive forces between molecules lower the molecules' energy, making them more energetically stable. But since molecules at the
liquid surface experience fewer attractions, they are less stabilized: water
at the surface has a higher energy than water deep in the bulk of the liquid.
In other words, there is an energetic price to be paid for creating an
interface between water and air: surfaces cost energy.

Like water flowing downhill, all physical systems have a tendency to
seek the state of lowest energy.* This means that, since surfaces have an

---

*Later I shall need to be more precise about this, in particular to specify that it is actually a
quantity called the *free energy* that systems tend to minimize. This principle of minimization of
free energy is what drives all processes of change. It is given a different expression in the
second law of thermodynamics, a concept that we will see to be central to the process of
pattern formation. Let me also mention now that I describe this minimization principle as a
'tendency' because it may be frustrated by obstacles, just as water can be kept from flowing
downhill by a dam wall.

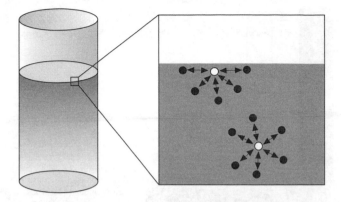

FIG. 2.15 Surface tension is a result of the asymmetrical attractive forces experienced by molecules at the liquid surface. Deep in the liquid, attractive forces on a molecule due to its neighbours are equal, on average, in all directions. But at the surface there is a net 'downwards' tug that tends to smooth away 'bumps' in the surface.

energy premium attached, there is a driving force to keep the surface area as small as it can be. A volume of liquid suspended in free space, like a water droplet in mist, will take on the form of a sphere because this is the shape with the smallest surface area. It amounts to the same thing to say that surface tension pulls the droplet into a spherical shape: surface tension and the energetic cost of a surface are equivalent expressions of the same basic fact that molecules at surfaces are less stabilized by attractive forces than those in the bulk.

It is not hard to see how surface tension produces the spherical form of liquid droplets. But in a cylindrical column of liquid it produces regular patterns. The droplet-inducing tug may make the column unstable, urging it to break up into blobs of equal size. The strands of spiders' webs are often coated with evenly spaced pearl-like beads of fly-catching glue (Fig. 2.16a). The spider has not painstakingly placed each of these blobs at carefully measured intervals; rather, it simply coats the thread in a continuous film of glue, which then fragments through the action of surface tension. The tiniest, random wobbles of the column's surface

59

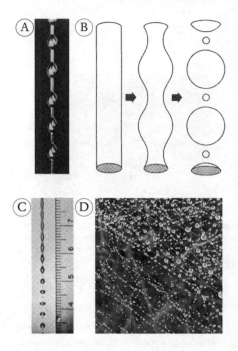

FIG. 2.16 A coating of 'glue' on the threads of a spider's web breaks up spontaneously into a string of pearl-like beads (a). This beading process, called the Rayleigh instability, is a fundamental property of a narrow cylindrical column of liquid, and it results in a particular droplet size (b). It can be seen also in the break-up of a narrow jet of water (c), and in the pearling of dew on a spider's web (d). (Photos: a, c, from Tritton, 1988. d, Olddanb)

become accentuated as surface tension pulls 'inwards' on the concave faces and draws the column into a series of narrow necks (Fig. 2.16b). This 'pearling' phenomenon is called the Rayleigh instability, since it was studied by the British scientist Lord Rayleigh (John William Strutt) at the end of the nineteenth century. What qualifies it as a patterning process, rather than mere fragmentation, is that the pearls are all of more or less equal size and spacing. That is because, although the instability acts on all undulating perturbations of the column, there is a certain wavelength that is the *most* unstable, and this determines the size and separation of the resulting droplets. In fact it is common for each pair of droplets in the broken-up column to be separated by a much smaller drop that forms in the narrow neck (Fig. 2.16b). That is not visible on a spider's web, where the intervening liquid just forms a uniform coating on the 'free' thread; but it can be seen when the Rayleigh instability disrupts a thin columnar jet of water (Fig. 2.16c). This instability also creates the glittering 'string of pearls' adornment of spider webs by early morning dew (Fig. 2.16d ).

## BALLOON GAMES

If surface tension pulls liquids into shapes of minimal area, then why do soap bubbles exist? Here is a liquid stretched into a thin film with a surface area far, far greater than that of a spherical droplet made from the same volume of fluid. What sustains it in this state, so apparently profligate with surface energy?

The answer is, as you might have guessed, the soap. You cannot blow a bubble like this from pure water;* but add a little soap or detergent and the liquid seems to forget its aversion to surface area. This is not because the surface of a soap film evades an energetic price, but because that cost is much lower. Soaps contain molecules called surfactants (a condensation of 'surface-active agents'), which gather at the water surface and greatly reduce the surface tension. That is quite the contrary to what our intuition tells us about soap bubbles somehow having a 'stronger skin' than water: it is indeed more stable, but only because the surface tension is reduced.

The key to surfactants is that they have split personalities: part of them is soluble in water, and part is insoluble. The soluble part is generally an electrically charged (ionic) 'head group', which is attached to an insoluble 'fatty' tail with a chemical structure like that of hydrocarbon oils and waxy greases. Molecules with this dual nature are called amphiphiles ('liking both'): in soaps and detergents, they are comprised of a water-loving (hydrophilic) part and a water-fearing (hydrophobic) part.

These amphiphilic molecules are most stable when their hydrophobic tails are out of the water. So they sit at the water surface with the head groups immersed and the tails protruding, creating a film one molecule thick that coats the surface (Fig. 2.17). (Another way to shield the tails from water is for the surfactants to clump together with the head groups

*The froth that sits on top of a pint of beer—a mass of fused bubbles—is stabilized by organic compounds in the beer, which do the same job as soap molecules. The same is true for the foam that coats the shoreline of a storm-tossed sea, stabilized by the secretions of marine organisms.

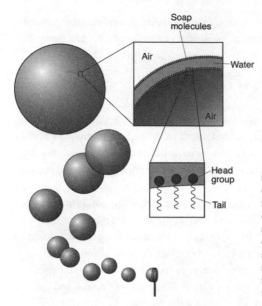

FIG. 2.17 The surface tension of the liquid in a soap bubble film is reduced by the presence of soap molecules at the surface. These molecules, examples of so-called surfactants, have a water-soluble head and a water-insoluble tail which pokes out from the water surface.

outermost and the tails buried inside. These structures, called micelles, have considerable pattern-forming potential, as we shall see.) A water surface coated by surfactants with their tails exposed is more stable than one where the surface is bare: it has a lower surface tension.

All the same, surface tension still pulls on a soap bubble and gives it a tendency to shrink. But the pressure of the air trapped inside opposes this shrinkage, and at some point the inward force of the surface tension is exactly balanced by the air pressure, and the bubble achieves its equilibrium size.

A soap bubble also feels the pull of gravity, which drags the water in the film towards the bottom of the bubble. This gradually thins the upper film until it reaches a point where it is too thin to remain stable: the film ruptures and the bubble pops. Interference of light reflected from the front and back faces of these films creates colours that depend on the film's thickness. These can give rise to swirling patterns of colour in soap bubbles, which turn silvery and then black as the film thins to breaking point.

The surface tension of a soap film may be relatively small, but it is not zero: there is still a driving force for the bubble to adopt the shape that minimizes the surface area while enclosing a given volume of gas. For a closed, isolated film, this 'minimal surface' is a sphere. But for films suspended on wire frames, the minimal surfaces can be complex and rather beautiful, with elegant, curving contours that can adapt themselves to any arrangement of the boundaries. Recognizing that soap films reveal how to cover an area with the greatest economy of material, the German architect Frei Otto has used them to design lightweight membrane structures. These tent-like shapes feature in many of Otto's designs, such as that for the Olympic stadium in Munich (Fig. 2.18a). To deduce what these shapes should be, Otto made wire models of the framework from which the membranes would hang, and then dipped them in a soap solution to drape the frame with a soap film (Fig. 2.18b). Calculating these minimal shapes mathematically is a challenging problem; soap films gave Otto an instant experimental answer.

## STACKING BUBBLES

When two soap bubbles come together, they coalesce and are flattened at their point of intersection. This partition is perfectly flat only if the two bubbles are of the same size, because the gas pressure inside a bubble depends on its radius: the smaller the radius, the greater the pressure. So a small bubble has a higher internal pressure than a larger one, and when they coalesce the partition between them will bulge out into the larger bubble (Fig. 2.19a).

How about three coalescing bubbles? The same applies: unequal sizes create bulges in the partitions, whereas three equal bubbles form a pleasingly symmetrical triad with flat internal walls that meet at angles of 120° (Fig. 2.19b). Now let's think about four bubbles. Will they similarly join into a cluster with square symmetry (Fig. 2.19c)? This probably looks unnatural to you, and that's because it is. Four equal bubbles arranged like this are rarer than four-leafed clover—in fact, they simply don't exist, because the bubbles will immediately rearrange themselves into a cluster like that in Fig. 2.19d, where no more than three bubbles meet at any point.

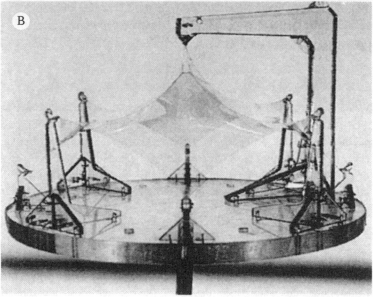

FIG. 2.18 The elegant area-minimizing shapes of soap films have inspired architects such as Frei Otto, whose used them to plan his design for the Olympic stadium in Munich in 1972 (a). Otto used soap films stretched across wire frames to map out the curves of his membrane structures (b). (Photos: a, Duncan Rawlinson; b, Michele Emmer, University of Rome 'La Sapienza'.)

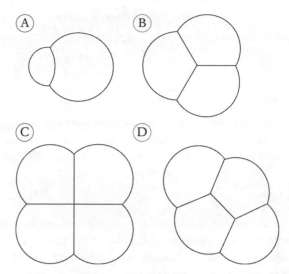

FIG. 2.19  How bubbles get together. When two bubbles of different sizes merge, the interface is convex on the side of the smaller one because of the difference in internal gas pressure (*a*). The interfaces between a trio of bubbles meet at angles of about 120° (*b*). With four bubbles, the intersection of four bubble walls (*c*) is unstable, and such a juncture will rearrange itself so that only three walls meet at any point (*d*).

D'Arcy Thompson noted that this is also what happens in clusters of living cells. 'In embryology', he said, 'when we examine a segmenting egg, of four (or more) segments, we find in like manner, in the majority of cases if not in all, that the same principle is still exemplified.' This isn't universally true, as Thompson admitted: four-celled human embryos do typically contain the kind of fourfold junctions that are unstable in soap bubbles, for example. This is a reminder that living cells are far more complex than soap bubbles. Their membranes are made of amphiphilic molecule called lipids; but whereas those in a soap film have their hydrophobic tails poking into the air, cell membranes are surrounded by watery fluid, and so the molecules arrange themselves the other way round, with the hydrophilic head groups pointing outwards on either side of the film and the hydrophobic tails shielded from water inside this double layer (Fig. 2.20). And these membranes are studded with all kinds of biomolecules that determine how the cell interacts with its

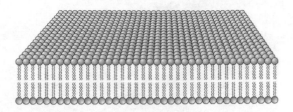

FIG. 2.20 Cell membranes are made from double layers of surfactant molecules called phospholipids. These 'bilayers' are studded with other membrane components, such as protein molecules, and are sometimes strengthened with a web of protein filaments called the cytoskeleton.

environment. There are, for example, protein molecules that control the way cells stick to one another so that they form delicately structured tissues and organs, rather than merely clustering like a mass of bubbles. In view of this, it is hardly surprising that clusters of living cells do not necessarily resemble groups of soap bubbles. What is surprising is that sometimes they *do*.

That can be seen, for example, in the cellular substructure of a fly's compound eye. Each lens-shaped facet is made from many cells, including a cluster of four light-sensitive cone cells. The cell biologists Takashi Hayashi and Richard Carthew have found that these clusters have the same arrangement as groups of four soap bubbles (Fig. 2.21*a*). In flies with a genetic mutation that gives them a variable number of cone cells, both larger and smaller clusters also show soap-bubble-like patterns (Fig. 2.21*b–e*). This suggests that, even though the binding of these cone cells to one another is controlled by adhesion proteins, the criterion that determines the cluster shapes is the same as that in soap bubbles: minimization of surface area.

If you look carefully at these soap-bubble clusters, you'll see that again there are junctions of no more than three bubble walls at any point. And where three walls meet, they do so at an angle of 120°, like the Mercedes symbol. It seems, then, that there are *rules* governing the way soap bubbles come together. These rules were first deduced at the end of the nineteenth century by the Belgian physicist Joseph Antoine Ferdinand Plateau. He reasoned that soap films meeting at junctions will always seek out the most mechanically stable arrangement, in which the forces

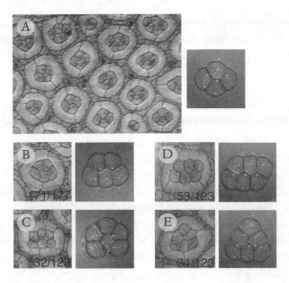

FIG. 2.21 *a*, Each facet of a fly's compound eye contains groups of four cone cells, arranged in the same way as soap bubbles (right). *b–e*, Some mutant flies have larger numbers of cone cells, which also adopt configurations found in the corresponding groups of bubbles. (Photos: From Hayashi and Carthew, 2004.)

acting on the films are all in balance. It is no trivial matter to calculate what that arrangement is, but Plateau did it—and he found that the criterion for mechanical stability is that the junctions should always be threefold, with 120° angles. Such junctions are now known as Plateau borders. The sharp curvature of the films in the 'corners' of a Plateau border means that the pressure inside the films is slightly lower here than in the flat regions, and so water is squeezed into the junction and thickens it. So most of the liquid in a conglomerate of bubbles resides at the Plateau borders where they meet.

Plateau's rules explain why a bubble raft of identical bubbles packs hexagonally: this is the way to make all the junctions between bubble walls threefold, with 120° angles. The result is a honeycomb-like array. That, then, is the optimal pattern for a flat array of identical bubbles. But what happens when bubbles are packed in three dimensions, as in a foam?

Engineers have long been interested in what determines the structure of a foam, since they have many practical uses: for example, to make lightweight, impact-resistant plastic packaging or insulation, or to smother fires, or to make a pint of beer look inviting. Foams are used in the living world, too: the spittle bug blows a froth to hide its larvae from predators.

A freshly blown foam may contain a lot of liquid, and the bubbles are a haphazard collection of roughly spherical cavities packed together with thick liquid walls between them (Fig. 2.22a). But over time the liquid drains away under gravity, and the walls become thinner. Then the foam becomes a three-dimensional network of junctions—Plateau borders—that divide up the space into a series of more or less flat-sided polyhedra (Fig. 2.22b). Plateau looked for the rules that govern these three-dimensional bubble arrays, and he found that when soap films meet at a vertex, there are always four of them—no more, no less. They meet at an angle of about 109.5°—the so-called 'tetrahedral' angle, which is the angle between lines running from the corners to the centre of a tetrahedron (Fig. 2.23). Again, Plateau calculated, this is the arrangement that guarantees mechanical stability of the bubble walls.

Plateau's rules explain why soap films held within geometrical wire frames spontaneously form structures of Platonic (Plateaunic?) simplicity and elegance (Fig. 2.24)—these correspond to the shapes that most closely obey Plateau's rules while conforming to the constraints imposed by the boundaries. Does a foam of very many bubbles share the geometric regularity of these arrangements? Let us suppose that we can construct a foam by carefully adding bubbles one at a time, each with identical size. What is the resulting geometry of the mesh of bubble polyhedra?

There are two factors at play here. We must look for polyhedra that satisfy Plateau's rules governing the angles of borders and vertices between bubbles. And the resulting cavities should minimize their total surface area. Is there a single well-defined way to partition space that meets these criteria?

That is the question Lord Kelvin (William Thomson) pondered as he lay in bed one September morning in 1887. It was just the kind of problem this great Victorian scientist loved to study: one that was easily posed, relevant in the everyday world, and amenable to experiment at home. By early November his niece Agnes came to visit, and she recorded that

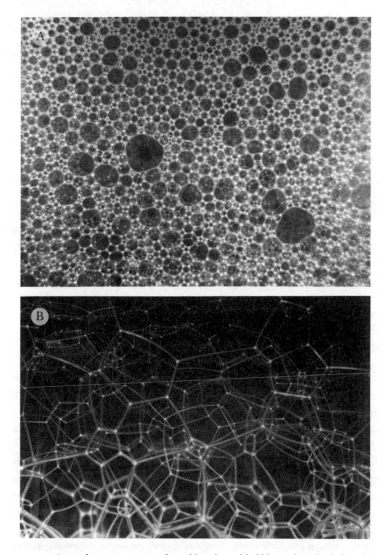

FIG. 2.22 A wet foam (a) consists of roughly spherical bubbles with walls thickened by water. As the water drains away under gravity, the bubbles become more polyhedral. This is a 'dry' foam (b). (Photos: Burkhard Prause, University of Notre Dame, Indiana.)

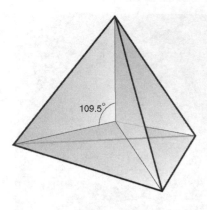

FIG. 2.23 Plateau deduced that precisely four soap films will meet at a vertex, with an angle of about 109.5° between them. This is the angle at the 'centre' of a tetrahedron.

> When I arrived here yesterday Uncle William and Aunt Fanny met me at the door, Uncle William armed with a vessel of soap and glycerine prepared for blowing soap bubbles, and a tray with a number of mathematical figures made of wire. These he dips into the soap mixture and a film forms or adheres to the wires very beautifully and perfectly regularly. With some scientific end in view he is studying these films.

It was on the previous evening that Kelvin wrote in his notebook what he suspected was the answer to the puzzle—that is, the shape of the cell that divides space into 'equal volumes of least partitional area' while observing (more or less) Plateau's rules. But before we see Kelvin's solution, we should recognize that he was not the first to consider this question.

In the eighteenth century the English clergyman Stephen Hales investigated the shapes that garden peas adopt when squeezed together: the answer, he said, was 'pretty regular Dodecahedra', for it seems he found many five-sided facets among the squashed peas (Fig. 2.25a). Perfect dodecahedra cannot be packed together perfectly to fill of the whole space—they leave little gaps. But the angles between edges are pretty close to the tetrahedral angle specified by Plateau: 116° between faces (rather than 120°), 108° between vertices (rather than 109.5°). And one need only distort the dodecahedron a little to close up the gaps. So Hales's answer looked like a fair guess.

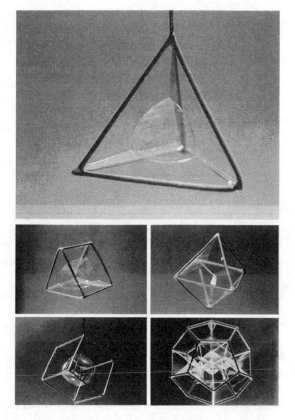

FIG. 2.24 Plateau's rules for the intersection of soap films are elegantly illustrated by films and bubbles held within geometric wire frames, which form regular, symmetrical shapes. (Photos: Michele Emmer, University of Rome 'La Sapienza'.)

But in 1753 the French zoologist George Louis Leclerc, Comte de Buffon, decided that Hales had meant that the cell shape was the *rhombic* dodecahedron, a twelve-sided figure with lozenge-shaped faces (Fig. 2.25b). This was accepted as the solution to the packing of squashed spheres until Kelvin published his new cell shape at the end of 1887. He said that the cell with minimal surface area was a 14-sided polyhedron made by

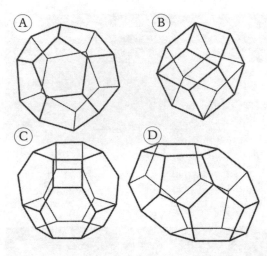

FIG. 2.25 Candidate cell shapes for a 'perfect' foam: the pentagonal dodecahedron (a), the rhombic dodecahedron (b), the truncated octahedron proposed by Lord Kelvin (c), and the beta-tetrakaidecahedron (d).

lopping off the vertices of a regular octahedron, giving six square and eight hexagonal faces. He called this a tetrakaidekahedron (Fig. 2.25c).

Kelvin's cell packs perfectly to fill space. But how well does it satisfy Plateau's rules? Where three edges meet in a vertex, two of these make angles of 120° while the third angle is 90°—in other words, both are rather different from the tetrahedral angle. But Kelvin showed that just a slight curvature of the hexagonal faces is sufficient to adapt all the vertices to the required 109.5°. Kelvin's solution was accepted for a long time as the most economical in surface area (D'Arcy Thompson mentions it admiringly), but Kelvin had no rigorous mathematical proof that it was the best. Thompson cautioned, however, that even if Kelvin's solution is the best in theory, in practice the polyhedra formed by squashing soft spheres could depend on the conditions under which they are squashed. He says that soft clay pellets will form shapes close to rhombic dodeca-hedra if compressed like Hales's peas, but that if they are first made wet so that they can slide over one another, they instead acquire square and hexagonal facets like those of Kelvin's tetrakaidekahedron. And he referred to experiments on the compression of lead balls by a botanist named J. W. Marvin, which apparently formed rhombic dodecahedra if first stacked in an orderly fashion like greengrocer's oranges but which formed irregular polyhedra with an average of 14 sides if poured into the press at random.

And what does all this imply for real foams? Do they obey Kelvin, or Buffon, or Hales, or none of them? When, in 1946, the botanist Edwin Matzke conducted a detailed study of cell shapes in foams carefully constructed from identical bubbles (called monodisperse foams), he found that none of the idealized, geometrical models quite fits the bill. Matzke's foams contained cells of many different shapes. Over half of the faces were five-sided, although only about 8 per cent of the cells came close to the shape of a regular dodecahedron. As for Kelvin's cell, only 10 per cent of the faces were four-sided, and none of the cells resembled the tetrakaidekahedron. Instead, most of them were rather like Marvin's randomly packed lead balls, with somewhat irregular shapes having an average of 14 sides each. Matzke proposed that these shapes were best approximated by a polyhedron with hexagonal and pentagonal faces called (in deference to Kelvin) a beta-tetrakaidecahedron (Fig. 2.25d). Matzke's results seemed to demolish the idea that even a monodisperse foam would have any regularity of cell shape.

All the same, that did not refute Kelvin's proposal for the cell shape that *in theory* minimized the surface area of a mechanically stable foam. And indeed it wasn't until a hundred years after Kelvin dabbled with trays of soapy water that a better solution was found. In 1993, physicists Dennis Weaire and Robert Phelan at Trinity College in Dublin discovered an even more economical way to pack polyhedral cells in an orderly fashion. It has to be said that their structure, while requiring less surface area than Kelvin's, is also considerably less elegant. Rather than invoking a single cell type with faces that are regular polygons, the foam described by Weaire and Phelan (Fig. 2.26) has a repeating unit made up from no fewer than eight polyhedra, six of which have 14 faces (two hexagons and 12 pentagons) and two of which have 12 (all pentagons). Only the hexagonal faces are regular (that is, having

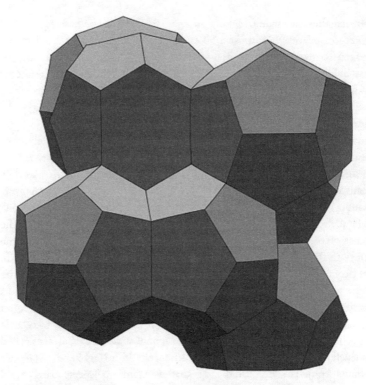

FIG. 2.26 This cellular structure, proposed by Denis Weaire and Robert Phelan, has a slightly smaller surface area than that made from Kelvin's cells, for the same enclosed volume. This entire unit is stacked together periodically in space, and consists of eight slightly irregular cells. (Image: Denis Weaire and Robert Phelan, Trinity College, Dublin.)

equal sides and angles); the pentagons have sides of different length and corners of different angles. Surprisingly, this complex figure can be stacked into a foam that has 0.3 per cent less surface area than a 'Kelvin foam' of the same volume, while maintaining Plateau's rules if the faces are very slightly curved.*

---

*John Sullivan at the University of Illinois at Urbana–Champaign, and Ruggero Gabbrielli at the University of Bath, have subsequently found other regular packings of polyhedra that improve on Kelvin's.

A team at the Australian architectural design company PTW stumbled over Weaire and Phelan's ideal foam while searching for a design for the swimming stadium for the 2008 Olympic Games in Beijing. The project was being orchestrated by the construction consultancy Arup, and the designers had decided that a building whose walls and roof were created from 'foam' could be lightweight, robust, translucent to daylight, and self-heating like a vast greenhouse. At first they explored Kelvin's foam, but found that its crystalline regularity lacked the 'organic' quality they were looking for. Weaire and Phelan's foam offered an appealing blend of order and irregularity, especially when the structure was rotated and then cut into slabs. This produced over a hundred different 'part bubbles'. Each of these cells was made from clear plastic film fitted over an ornate mesh of steel struts (Fig. 2.27). The resulting building, called the Beijing Water Cube, is one of the most remarkable products of the awesome building campaign motivated by the Chinese Olympics. In principle, this design also minimizes the amount of material needed to make the foam cells—but a saving of just 0.3 per cent at best hardly compensates for the complexity of fabrication and assembly, and it is probably best to see this as a triumph of aesthetics more than economy. Indeed, PTW designer Chris Bosse was so taken with the visual qualities of the new foam structure that he recreated it with students in an art installation in Sydney (Fig. 2.28).

Yet it seems hard to believe that a real foam would find a solution this complex. Weaire and Phelan decided to check. They developed a simple method for making a monodisperse foam that involved little more than the 'drinking straw' technique of blowing bubbles underwater, and they found that the foams generated this way were not necessarily as irregular and disordered as those Matzke had reported. In parts of the foam close to the walls of the vessel they often observed regular cells with square and hexagonal faces, like Kelvin's polyhedra (Fig. 2.29a). But deeper within the foam they found cells with hexagonal and pentagonal faces that fitted together in a manner much like that found in their 'minimal foam' (Fig. 2.29b,c). Perhaps, if the conditions are right, then, foams do indeed find something close to the Platonic ideal that economizes on surface while achieving mechanical stability.

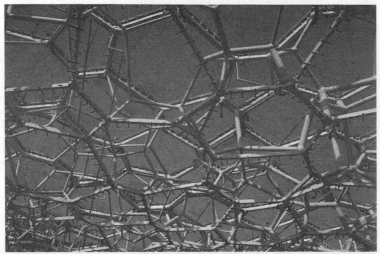

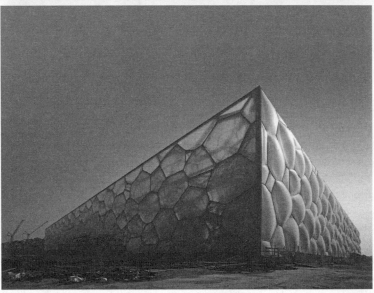

FIG. 2.27 The framework and the finished structure of the Olympic swimming stadium in Beijing for the 2008 Olympics (the Water Cube), based on Weaire and Phelan's ideal foam. (Photos: Ben McMillan.)

FIG. 2.28 The Weaire and Phelan foam has been assembled into an installation for an architectural show in Sydney. (Photo: Chris Bosse, PTW Architects, Sydney.)

## FACE TO FACE

Bees face an easier challenge in making their 'wax foam', because it is essentially just two-dimensional: the cells are uniform prisms with constant cross-section. But even here, there is no mathematical proof that the

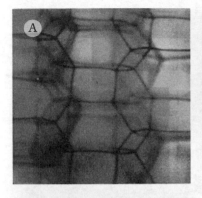

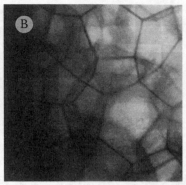

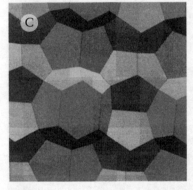

FIG. 2.29 What does a real dry 'ideal' foam look like? At its boundaries, the cells look like Kelvin's (a), but deeper inside they resemble those of the 'minimal foam' of Weaire and Phelan (b, c). (Photos and image: Denis Weaire and Robert Phelan, Trinity College, Dublin.)

hexagonal arrangement of cells is the one that minimizes the wall area. If there exists a more complex cell shape that does fractionally better, none has been found for bubble rafts.

The challenge the bees face is not, however, quite that simple. A honeycomb consists of *two* arrays of hexagonal cells married back to back, and the question then arises of how best to join the layers. This is a three-dimensional problem, and the most economical solution is not obvious. Honey-bees adopt a rather sophisticated structure in which each cell ends in a cap made from three lozenge-shaped faces (Fig. 2.30a). These are the components of a rhombic dodecahedron, and cells married with such end caps have a zigzag cross-section (Fig. 2.30b). Is that, then, the way to be most frugal with wax?

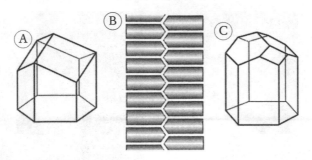

FIG. 2.30 The ends of a honeycomb's cells are fragments of rhombic dodecahedra, made up of three rhombic faces (a). The two layers of cells with these end caps marry up with a zigzag cross-section (b). Is this the minimal solution? A smaller surface area is obtained for end caps that are fragments of Kelvin's truncated octahedral (c).

Réaumur considered this question in the eighteenth century. Observing that bees make end caps of rhombuses with edges of equal length, he wanted to know the angles of these polygons that minimized the surface area. The maths was beyond him, so he asked the Swiss mathematician Samuel Koenig to solve the puzzle. Koenig showed that the ideal angles are about 109.5° and 70.5°, which are those seen in a regular rhombic dodecahedron and are also those observed in real honeycombs. To find this answer, Koenig needed to use the methods of calculus devised in the seventeenth century by Isaac Newton and Gottfried Wilhelm Leibniz. How on earth could the bees 'know' about that piece of new mathematics? The secretary of the French Academy of Sciences, Bernard de Fontenelle, could not believe that bees were capable of calculus—for that, he said, would surely mean that 'in the end these Bees would know too much, and their exceeding glory would be their own ruin.' Thus, he said, it must be that these mathematical principles were exercised by the insects according to 'divine guidance and command'. Darwin removed the need for such heavenly intervention by supposing that selective pressure would drive the bees to find the optimal solution by trial and error.

But are the bees' cell caps really optimal? By imposing the constraint of using three identical rhombuses, Réaumur ruled out the possibility of finding other geometries that did better. In 1964 the Hungarian

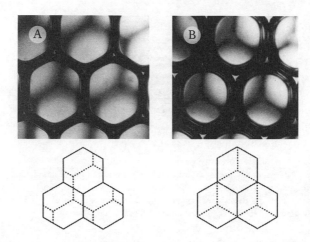

FIG. 2.31 Tóth's structures can be seen at the interface of a double layer of hexagonal bubbles (a). But if the bubbles contain more liquid in their walls, the faces at the interface change to rhombuses (b), giving a junction more like that seen in real honeycombs. (Photos: Denis Weaire and Robert Phelan, Trinity College, Dublin.)

mathematician László Fejes Tóth broadened the range of possible solutions and found that a more elaborate cap structure made from square and hexagonal facets has fractionally less surface area (Fig. 2.30c). Just as the rhombic cap is a fragment of the rhombic dodecahedron, so Tóth's cap is a piece of Kelvin's tetrakaidekahedron. Tóth admitted, however, that as his cap was more complicated, there was no guarantee that it was biologically superior, for the bees might have to expend more effort in making it.

Weaire and Phelan have put these conjectures to the test by creating double layers of hexagonally packed bubbles in the space between two glass plates, mimicking the cell structure of the honeycomb. They found that the interface between the two layers of bubbles does adopt Tóth's structure, which can be identified from the pattern made by the junction of bubbles in projection (Fig. 2.31a). But if the bubble walls are thickened by adding more liquid, the interface suddenly switches to a different configuration: the three-rhombus pattern seen in real honeycombs (Fig. 2.31b). This switch is apparently caused by an altered balance in surface energies as the bubble walls get wider and more curved. So perhaps bees really do make the most efficient structure, under the circumstances?

D'Arcy Thompson was happy to believe that indeed bubble mechanics—the play of surface forces—governs the shape of a honeycomb, pulling the cells into two hexagonal arrays that join back to back with minimal surface area. In that case, there is no need for any Darwinian 'just-so' story about the bees having evolved to make economies of wax: the economy comes instead purely from mechanical forces and surface energies, which also create a rather splendid, orderly natural pattern. 'The bees make no economies', he insisted, 'and whatever economies lie in the theoretical construction, the bee's handiwork is not fine nor accurate enough to take advantage of them'.

Researchers in Germany and South Africa, led by Christian Pirk of the University of Würzburg, have recently shown that Thompson's hypothesis works in principle. They poured molten bees' wax into trays containing cylindrical rubber bungs arranged in hexagonal arrays. The wax initially filled the space between the bungs completely, producing a series of cylindrical holes separated by thin wax walls. But as it cooled and hardened, surface tension pulled the walls into a hexagonal shape, flat along the edges and thickened at the corners, much like real honeycombs. Pirk and colleagues think that this may be just what happens in the honeycomb, where the bees themselves act as the roughly cylindrical moulds, their bodies packed together in a hexagonally ordered fashion and their body warmth softening the wax to the point where it can flow.

But it's not clear that the hive gets quite warm enough for this—and in any event, it's hard to see why the bodies of swarming creatures should become ordered into a perfect hexagonal array that acts as the mould. It seems more likely that D'Arcy Thompson's reliance on surface tension for this astonishingly regular pattern deeply underestimates the capabilities of the female worker bees who construct the comb. Rather than relying on physics to do the job spontaneously, building a comb looks like painstaking work. The worker secretes flakes of wax from glands on the underside of her abdomen, which she pulls out using a hind leg and then chews to make it soft and mouldable. Each softened flake is put in place one by one, and so the comb is made much as we would construct a brick wall. And to create walls that meet at precisely 120°, the worker uses a set of physio-logical measuring tools that are still imperfectly understood. She can sense the vertical direction defined by gravity by using her head as a plumb line, its orientation gauged by an organ on her neck. It used to be thought that this vertical helped to define the orientation of the rows of cells, so that two opposite sides of each cells sit either vertically or horizontally. But recent experiments seem to show that the orientation of the cells is defined instead by the direction of the surface on which the first row of cells is constructed. Thus the bees seem able to measure 120° angles with respect to any arbitrary direction.

They do more than that. The cells also tilt upwards slightly along their axis at an angle of about 13° with respect to the horizontal, which prevents the viscous honey from running out. The thickness of the cell walls is machined to an incredibly fine tolerance of two-thousandths of a millimetre, thanks to tactile organs that enable the bee to gauge this thickness from the wall's flexibility; and what is truly extraordinary is that each comb emerges as a regular array despite being the product of many workers, without the mismatches that would be expected from tilers each beginning their work at a different point. Even the comb as a whole is aligned to the Earth's magnetic field, so that many-layered combs stack efficiently even when produced in near-darkness. One way or another, then, bees possess a set of finely honed tools that they acquire through genetic inheritance, and it is possible that these tools would enable the insects to make honeycombs with quite different patterns if their genetically programmed instincts permitted it. That they do not is surely a mathematical exigency: the hexagonal form of a honeycomb *does* follow from the dictates of geometry on packing efficiency and minimization of surface, even if we cannot be sure that it represents the totally optimal solution from a mathematical point of view. Yet D'Arcy Thompson was wrong to assume that the bees did not need to conduct any hand-tooling or careful measuring to realize that structure, but could rely merely on physical organizing forces.

Although Thompson was a little too enthusiastic here in his determin-ation to find physical forces guiding biological structures, there is still a sense in which he was right to challenge the easy answers of Darwinism—or, in its modern incarnation, of genetics. For you can read the genome of the honey-bee from end to end, and will find nowhere within it the blueprint for a honeycomb. That is something which emerges only when the organism goes about her job—and not just the organism, but the whole hive, whose hexagonal storehouse is a collective effort. Sometimes you need to take a step back, rather than peering closer, in order to make sense of what you are seeing. D'Arcy Thompson would surely have no argument with that.

## CHEAP WAYS TO FILL UP SPACE

As Frei Otto appreciated, finding a surface of minimal area between any given set of boundaries could hardly be simpler: you make the boundaries from wire and dip them in a soap solution. But, mathematically, the problem is rather fearsome. Leonhard Euler was the first to make significant progress on it, when he showed in 1744 that the surface of least area spanning two coaxial circular hoops is a vase-shaped form called a catenoid (Fig. 2.32). This is the shape of the meniscus that adheres to your finger when you dip it in water and then withdraw it slowly. Joseph Lagrange studied these 'minimal surfaces' using calculus in the 1760s, but it wasn't until 1776 that the French mathematician Jean Meusnier (who was also a military general and an expert on balloon aeronautics)

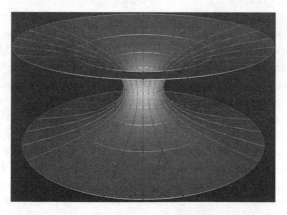

FIG. 2.32  The minimal surface spanning two coaxial circular hoops is called a catenoid. (Image: Matthias Weber, Indiana University.)

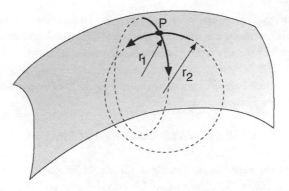

FIG. 2.33    The mean curvature of a surface at a point P is the sum of the curvatures in two perpendicular directions at that point, which are calculated from the radii of the circles that match the surface contours at that point.

identified the defining property of such a surface. The fundamental characteristic is not just that the shape minimizes the surface area but that this surface has *zero mean curvature* at every point. The curvature of a surface is related to its radius: the smaller the radius, the larger the curvature. For a surface that curves in two dimensions, like a hill or a bowl, the mean curvature is obtained by adding the curvatures in two perpendicular directions (Fig. 2.33). For a shape like Euler's catenoid, the surface curves one way (hill-like or 'downwards') in one direction and the opposite way (bowl-like or 'upwards') in the perpendicular direction. Thus, one curvature has a positive value, and the other is negative. Any portion of such a surface is shaped like a saddle. For a surface of zero mean curvature, these two values of curvature perfectly cancel each other out everywhere. This might seem odd, for the catenoid is clearly *highly* curved—but the *mean* curvature is nevertheless zero. If we think in terms of soap films, we saw earlier that the pressure difference between two sides of a film is related to its curvature: the film is only flat when there is no difference in pressure. A shape like the catenoid is not flat, but nonetheless its zero mean curvature means that the pressure is equal on both sides.

In 1834 the mathematician Heinrich Scherk discovered that it is possible to make minimal surfaces that have no boundaries: they can extend through space for ever. Scherk showed that a saddle-shaped 'brick' could be joined to other identical units at its boundaries to make a lattice that repeats indefinitely (Fig. 2.34). This is called a *periodic minimal surface*, and it has the property of dividing three-dimensional space into two interpenetrating but independent labyrinths: you cannot move from one labyrinth to the other without passing through the surface between them. For this reason, the structure is said to be *bicontinuous*.

Another German,* Hermann Schwarz, subsequently found another bicontinous periodic minimal surface while studying a problem posed by Plateau: what is the shape of a soap film stretched within a tetrahedral framework and touching all four corners? The saddle-shaped solution

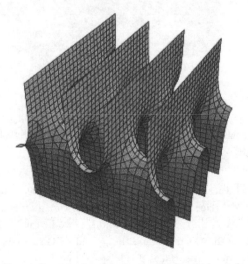

FIG. 2.34   Scherk's periodic minimal surface. (Image: Matthias Weber, Indiana University.)

*Lest national pride should be offended, let me point out that both Scherk and Schwarz were born in Germanic states that are now part of Poland.

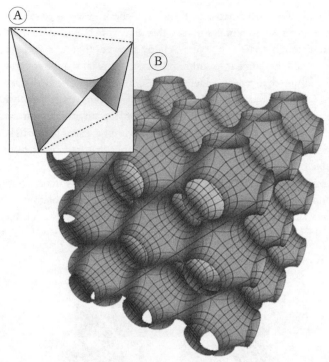

FIG. 2.35  Schwarz's minimal surface spanning the four corners of a tetrahedron (*a*) can be used as the 'building block' for a periodic minimal surface called the P-surface (*b*). (Image *b*: Matthias Weber, Indiana University.)

deduced by Schwarz (Fig. 2.35*a*) can, like Scherk's, be pieced together into the so-called P-surface (Fig. 2.35*b*). The American mathematician Alan Schoen found two other simple bicontinous periodic minimal surfaces in the 1960s, called the D-surface and the gyroid or G-surface (Plate 1). All these periodic minimal surfaces divide space into two regions that have everywhere an equal pressure on both sides.

Because they represent surfaces of minimal area, periodic minimal surfaces are in principle energetically favourable ways of weaving a membrane throughout space. Does nature have any use for that particular economy? Most certainly it does.

## CRYSTAL CELLS

Cell membranes, as we noted above, are a little like inside-out soap films: double layers of amphiphilic molecules called lipids, back to back and held together by the aversion of the lipid tails to water. Similar bilayers can be made artificially from surfactants or lipids dissolved in water. In low concentrations the molecules just float around on their own, or gather at the water surface with tails in the air. Increase the concentration and they form globular clusters called micelles, with the tails innermost and the water-soluble heads at the surface. At still higher concentrations the surfactants gather into bilayers, which may close up into bubble-like compartments called vesicles or stack up in sheets. These sheets are fairly flexible, but they do not bend without an energy penalty, because the curvature prises the head groups of the outer layer apart and increases the exposure of the tails to water. This means that the structures adopted by bilayers are determined not just by their surface area but also by the amount of bending they entail. The balance can be rather subtle, and can give rise to vesicles with quite complex shapes, such as stars or doughnuts (Fig. 2.36). It is rather surprising that such shapes in these artificial cell-like bodies, which give them a super-ficial appearance of amoeboid life, can arise simply from the physical forces that determine form.

Bilayer sheets are agitated by little ripples, a consequence of the random motions that heat induces in molecules. In a stack of bilayer sheets, called a lamellar phase, these fluctuations can make adjacent sheets touch and fuse, rather as if they were two soap films, opening up channels and pores connecting regions of space that were previously separated (Fig. 2.37). The pores introduce curvature, and so they cost energy. But if there is enough energy around in the thermal motions of the sheets to pay that price, they may merge into a web of tunnels. These can be arranged in a disorderly, haphazard manner, giving the stack of membranes the random, perforated structure of a sponge. The result is a so-called sponge phase, or more figuratively, a 'plumber's nightmare'. But it is also possible for the pores to be arranged in an orderly pattern. Why would that happen? If two pores are situated close together, that increases the curvature in the region between them, which costs more energy than forming the two

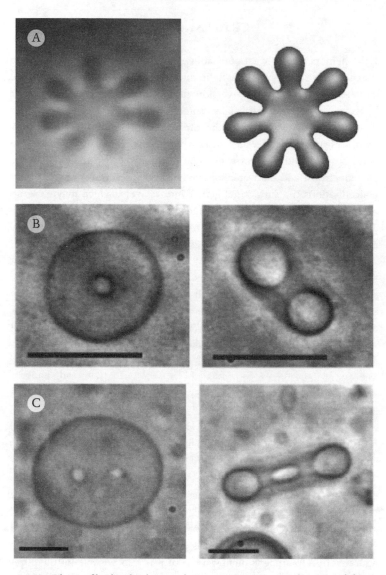

FIG. 2.36 Bilayers of lipid molecules may close up into some surprisingly contorted shapes, such as starfish (a), doughnut (b), and double-doughnut (c), vesicles. Both top and side views are shown in b and c. (Photos: a, Udo Seifert, Max Planck Institute for Colloid Science, Teltow-Seehof. b, c, Xavier Michalet and David Bensimon, Ecole Normale Supérieure, Paris.)

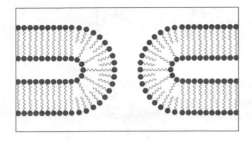

FIG. 2.37 Pores can form spontaneously between adjacent bilayer sheets in the so-called lamellar phase, providing channels between previously separated regions of space.

pores in isolation. So the pores tend to stay apart—in effect they repel one another. If there are lots of pores in the system, their repulsion leads them to sit a more or less even distance apart, creating a sort of 'tubular crystal'. A slice through one such ordered pattern (Fig. 2.38) shows that this is a bicontinuous structure that divides up the space into two interpenetrating regions.

Another of these ordered bicontinuous phases formed from surfactants, discovered in the 1960s by the Italian chemist Vittorio Luzzati, has essentially the form of Schwarz's P-surface (Fig. 2.35), and is called the cubic P-phase, since the basic repeating unit is one with the symmetry properties of a cube. Here the pores form in all three dimensions, creating not distinct layers connected by channels but instead a labyrinthine network—in fact, two interpenetrating networks—laced through all of space.

The key to the formation of periodic minimal surfaces by surfactants is that, with the surface area of the bilayers fixed by how much surfactant there is in the solution, these surfaces provide the best way to pack that surface into the available volume: there are no highly bent regions, since the mean curvature is zero everywhere.

Luzzati wondered whether, given that cell membranes are so similar to surfactant bilayers, structures like these might be found in living cells, where they might perhaps provide complex plumbing systems. Indeed, the tangle of membrane channels and pores in our cells known as the smooth endoplasmic reticulum, where lipids and some proteins are manufactured, looks very much like the disordered sponge phase (Fig. 2.39). Could this become an ordered network?

FIG. 2.38  In a bicontinuous phase of surfactant bilayers, the space is divided into two distinct networks that interpenetrate without being interconnected. Here I show a slice through a simple bicontinuous phase made up from pores connecting adjacent bilayers. The two subspaces are shown in light and dark grey.

As we have seen, real cell membranes are considerably more complex than surfactant films or soap bubbles—in particular, they contain mixtures of several different lipids (which means that their surface tension may vary from place to place) and they are studded with proteins and other biomolecules which affect the propensity of membranes to bend and adhere to one another. So it is not obvious that physical forces such as surface minimization and curvature energy will be the only, or even the primary, factors determining membrane shape.

Nonetheless, it appears that sometimes they are. In 1965 the plant biologist Brian Gunning, working at the Australian National University in Canberra, saw curious regular patterns in electron microscope images of plant cells (Fig. 2.40a). The dark regions in these images trace out shadow-like projections of membranes, which are here apparently ordered

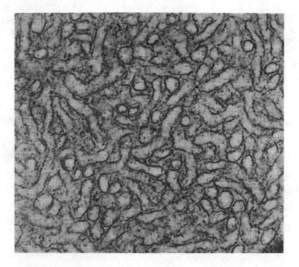

FIG. 2.39   The cells' smooth endoplasmic reticulum is a disordered 'sponge' of membranes perforated by pores. (Photo: Don Fawcett.)

in regimented compartments or channels. Gunning proposed that these structures, called prolamellar membranes, have the same form as the cubic P-phase. Ten years later, he and botanist Martin Steer pointed out the connection between these 'crystalline' prolamellar membranes and the forms exhibited by soap films, including the periodic minimal surfaces that could be constructed from repeating units. Then in 1980 Kåre Larssen at Lund University in Sweden and his co-workers compared some of Gunning's microscope images with the patterns predicted theoretically for membranes shaped like Schwarz's periodic D-surface (Plate 1). The Lund team searched through the literature of cell biology and discovered many other orderly membrane structures, which they called cubic membranes, in organisms ranging from bacteria to rats to lampreys (Fig. 2.40b,c). They showed that many of these corresponded to periodic minimal surfaces or closely related structures.* They are common, said the Swedish

---

*Some of these structures correspond to periodic surfaces of *constant* mean curvature—that is, not surfaces with *zero* mean curvature at all points, but ones for which the curvature is non-zero but identical everywhere.

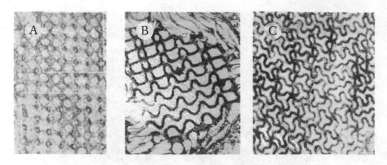

FIG. 2.40   Periodic membrane structures are common in living cells. Many of these appear to be related to periodic minimal surfaces: the P-surface in leaf membranes (a), the D-surface in algae (b) and the G-surface (gyroid) in lamprey epithelial cells (c). (Photos: a, from Gunning, 1965; b, from McLean and Pessoney, 1970. All images kindly provided by Tomas Landh, Lund University.)

researchers, in the endoplasmic reticulum and in the membranes of mito-chondria (cell compartments that generate metabolic energy) and lyso-somes (compartments that break down proteins and lipids).

Now, we cannot assume that, just because biological cubic membranes have the same structures as those of the periodic surfaces seen in simple surfactant bilayers, they are formed for the same reasons. For one thing, the cell structures tend to be several times bigger. Also, the cell is too restless a thing to be sure that its structures correspond to those of lowest energy. All the same, these cubic membranes are surely formed spontan-eously, moulded by physical forces rather than laboriously built piece by piece like the bee's honeycomb. And since they are so common, it is safe to assume that nature has found some use for them. No one yet knows what their purpose is. It is not hard to imagine, however, why a cell might find it expedient to divide up space this way, creating two networks that permeate space while remaining isolated from one another. Maybe, for instance, this suppresses 'cross-talk' between two biochemical processes. Making cubic membranes is also a good way of creating a lot of 'work surface' in a small volume: the processes of protein synthesis in the endoplasmic reticulum happen at membrane surfaces. And creating regularly spaced compartments might improve the efficiency of assembly-line molecular processes. But these are just guesses.

However, there seems to be a very clear function for the natural bicontinuous structures that have been discovered in the wings of butterflies. Doekele Stavenga of the University of Groningen in the Netherlands and his collaborator Kristel Michielsen found that some butterfly wing scales—tiny overlapping flakes of hard cuticle arranged like roofing tiles on the wing surface—contain porous labyrinths in the shape of the gyroid or G-surface (Fig. 2.41). They investigated the wing scales of several species of papilionid and lycaenid butterflies, including the European green hairstreak (*Callophryus rubi*), the emerald-patched cattleheart (*Parides sesostris*) and the Indian Kaiser-i-Hind butterfly (*Teinopalpus imperialis*), under an electron microscope. The scales are perforated with what appears to be a regular lattice of holes; on close inspection, the holes prove to be interconnected in a way that matches the patterns expected for a bicontinuous gyroid

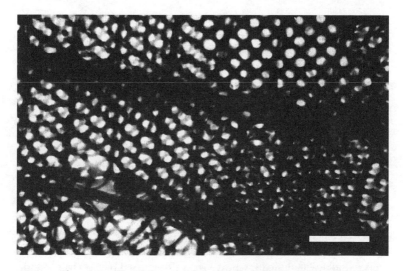

FIG. 2.41 The wing scales of the green hairstreak butterfly *C. rubi* are made from a lattice of hard cuticle with the structure of the gyroid periodic minimal surface, seen here in the electron microscope. The white bar corresponds to a length of one thousandth of a millimetre. (Photo: from Ghiradella and Radigan, 1976.)

structure. Stavenga and Michielsen think that this is a remnant of such a structure in the endoplasmic reticulum of the cells that form the wing scales, which gets covered in cuticle (mostly a tough biological material called chitin) as the scale develops. Eventually the cells die, leaving just a hardened husk of microscopically patterned cuticle.

But why? Materials imbued with patterns like these, whose features have a size comparable to the wavelength of visible light, can interact with light to produce striking optical effects. Many butterfly scales are covered in arrays of ridges that scatter light and generate bright, iridescent colours. The gyroid structure behaves similarly: in *P. sesostris* and *C. rubi*, it creates the characteristic green colour of the wings. So it seems that this microscopic patterning process is used here to create the basic building blocks of the spectacular, larger-scale patterns of butterfly wing markings, whose genesis is traced in Chapter 4.

## PATTERNED PLASTICS

This 'freezing' of periodic minimal surfaces to make robust labyrinths from materials much tougher than soap films or cell membranes has been achieved artificially by using synthetic polymers. These, the fabric of plastics, contain large molecules consisting of many smaller molecular units linked together by chemical bonds, typically into simple chains. In general, different polymers do not like mixing: throw together two liquid polymers and they will separate into two layers, like oil and water, with the densest on the bottom. But what happens if you shackle two different chains together so they cannot escape one another? These hybrids are called block co-polymers: chains divided into blocks of different type. Chains of two blocks—diblock co-polymers—are thus rather like elongated surfactants, with a head and a tail (or in this case, two tails) that have contrasting 'personalities'.

Block co-polymers accommodate this conflict of preferences by finding a compromise between mixing and separation. Blocks of the same type clump together in domains whose size is determined by the chain length. This can lead to the formation of roughly spherical, microscopic blobs of one polymer type surrounded by a 'sea' of the other (Fig. 2.42). The rubbery material used for the soles of training shoes has this structure. It

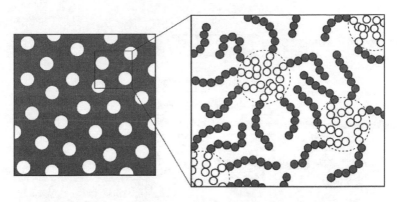

FIG. 2.42 Block co-polymers contain molecular chains with chemically different segments, which separate out into distinct domains.

is in fact a *triblock* co-polymer in which a segment of the polymer polybutadiene is sandwiched between two blocks of polystyrene. This material is 'thermosetting': unlike real rubber, it can be repeatedly melted and then reset into its rubbery form, the microscopic blobs fragmenting and reforming.

This kind of partial separation is not a bad compromise, but it creates a lot of surface between the domains of different blocks. There's no getting around this, however—if you're going to have some segregation, you have to make some surface. The question is then how best to arrange the surface so as to minimize its area. This depends on the relative length of the segments: as the length ratio changes, a diblock co-polymer can form a variety of different patterned structures (Fig. 2.43). If one block is short, spherical blobs are the best option; and if there are many of these, they can become ordered into a regular array, like a raft of bubbles (Fig. 2.44a). If the shorter blocks are a little longer, the spheres become cylinders (Fig. 2.44b). For blocks of roughly equal length, the chains form flat sheets, analogous to the lamellar phase of surfactants. But in between these cylindrical and lamellar arrangements, some block co-polymers adopt arrangements corresponding to bicontinuous periodic minimal surfaces, such as the gyroid phase (Fig. 2.44c). Scherk's minimal surfaces (Fig. 2.34) have also been seen in co-polymers. Triblock co-polymers can adopt even more complex

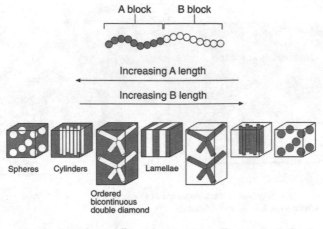

FIG. 2.43 Co-polymers with two different chain segments will spontaneously form a variety of ordered structures, depending on the ratio of chain lengths.

periodic structures, with textures that resemble those of woven fabrics (Fig. 2.44d).

It is not clear that these patterns always represent mathematically ideal minimal surfaces. For one thing, if the two blocks are of different length then the interpenetrating networks are not completely equivalent. And the patterns are also influenced by the packing and stretching of the polymer chains, which impose energy costs that have to be folded into the balance. But that does not undermine the fact that these spontaneous patterns result from a delicate interplay of forces, and that they can be altered dramatically in scale or structure by a small shift in that balance. These patterns cannot be predicted simply by considering the individual building blocks or trying to work out how a few of them might be stacked together. They are *emergent* properties of the whole system.

## FOSSIL FOAMS

Now we are ready at last to return to the fantastic mineral meshes of Ernst Haeckel's radiolarian atlas. For we can see at once that the reticulated exoskeleton frameworks of *Aulonia hexagona* and its ilk resemble nothing

FIG. 2.44 Microscopic cross-sections of some of the ordered patterns formed by block co-polymers: an hexagonal array of spherical domains (a), and hexagonal array of cylinders (b, here seen partly face-on), a minimal-surface gyroid phase (c) and a more complex structure formed from chains with three segments (d). (Photos: Edwin Thomas, Massachusetts Institute of Technology.)

other than fossilized foams. And that, according to D'Arcy Thompson, is precisely what they are. To create this kind of web, he said, the organism blows bubbles and then sets their junctions in stone—that is to say, the microscopic marine plants and animals surround themselves with a layer of bubble-like vesicles (called vacuoles) made from organic membranes, and then allow a mineral, silica or calcium carbonate, to precipitate from the solution trapped at the Plateau borders where vacuole walls meet.

In some cases this hypothesis seems to fit very nicely with the biological facts. To make the three-pointed silica stars called spicules produced by sponges, a Plateau vertex provides the perfect mould (Fig. 2.45a). The simple exoskeleton of a silicoflagellate seems to obey Plateau's rules even down to the slight curvature of the struts owing to the pressure differences between 'bubbles' of different size (Fig. 2.45b). Haeckel's nassellarian skeleton looks just like a bubble suspended within a tetrahedral frame (Fig. 2.46a), while his aptly named *Lithocubus geometricus* is a bubble within a cube, and *Prismatium tripodium* a bubble within a trigonal prism (Fig. 2.46b,c).

Even more suggestive is the skeleton of the sea urchin *Cidaris rugosa*—a mesh of calcite (Fig. 2.47) that looks just like the cubic P-surface (Fig. 2.38). It seems a fair guess that this structure, like those in butterfly wing scales, was deposited on a template of soft organic tissue patterned along the same lines as the cubic membranes of cells. It is a good engineering design, giving strength and rigidity in all directions while using relatively little material: the skeletons are stronger than reinforced concrete.

How does an organism cast a foam in stone? Both radiolarians and diatoms are now known to secrete bubble-like structures called areolar vesicles, which remain attached to the organism's membrane wall (the plasmalemma) and pack into regular, foam-like arrays. A series of thin, tubular membranes thread through this foam, carrying a silica-rich solution which produces a mineralized mesh on the template formed by the areolar vesicles (Fig. 2.48). In some organisms, especially diatoms, a process like this may be conducted in a hierarchical manner over several scales, creating patterns within patterns. Thus, although the process by which the hard material is confined and deposited is controlled quite closely by 'biological design', nonetheless D'Arcy Thompson's basic notion of a mineralized foam seems to be sound.

There are biological advantages to these patterned mineral trellises. They offer robust protection at a low cost in materials and weight. Their open, porous structure leaves room for organic fibres and tissues to thread through, just as cells and blood vessels pass through the mineral fabric of bone. And there are more exotic benefits to the periodic patterning too. As we saw for butterfly wings, it can supply a kind of lattice for scattering light and creating colour. But orderly microscopic labyrinths have still

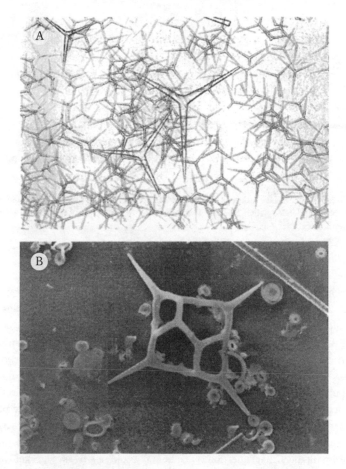

FIG. 2.45 The spicules of sponges appear to be mineral 'casts' of Plateau borders between groups of bubble-like vesicles (a). Plateau junctions are clearly evident in the exoskeletons of silicoflagellates (b). (Photos: a, Michelle Kelly-Borges, Natural History Museum, London; b, Stephen Mann, University of Bristol.)

more subtle and remarkable ways of interacting with light: they don't just scatter it, but can actually capture, confine, and guide it. By such means, the silica spicules of the deep-sea sponge *Euplectella* act as natural optical fibres that channel light.

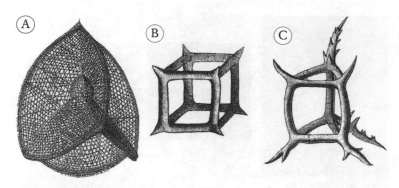

FIG. 2.46 Some of the exoskeletons drawn by Ernst Haeckel have the structures expected of bubbles suspended within cages: here, a tetrahedral (*a*), cubic (*b*) and trigonal prismatic (*c*) framework. (See Fig. 2.24).

Some researchers are interested in capturing these light-channelling properties in synthetic materials patterned in the same way, which might find uses in optical technologies such as telecommunications. They have used patterned biominerals such as the exoskeletons of diatoms and sea urchins as moulds, around which they cast webs and meshes of other inorganic solids. But if this sort of biological patterning really is basically the result of simple physical and chemical forces, rather than requiring a detailed blueprint encoded in an organism's DNA and being reliant on a protein work force to put the elements in place, might those same forces be harnessed in the laboratory to create wholly artificial materials that are spontaneously patterned? Can nature's strategies be adapted to make artificial radiolarians in the test tube?

Remarkably, that question was being asked even as Haeckel was discovering the astonishing diversity of shape and form in marine microorganisms. In the 1870s, the Dutch zoologist Pieter Harting wondered whether the 'calcareous formations' that Haeckel, Huxley, and others had reported might be reproduced by chemical means alone. He carried out experiments in which he crystallized calcium carbonate and phosphate from mixtures of ingredients that sound positively Shakespearean: egg white, gelatin, and, as he put it, 'blood, bile, mucus...and the liquor obtained by triturating chopped-up oysters in a mortar'. Out of this

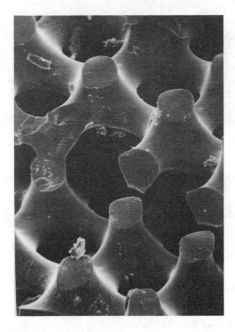

FIG. 2.47 The calcite skeleton of the sea urchin *Cidaris rugosa* appears to be a mineralized cast of the periodic P-surface. (Photo: Hans-Udde Nissen, kindly supplied by Michele Emmer.)

witches' brew came forth 'a considerable number of forms . . . which are, for the most part, found in nature'.

In particular, Harting described patterned spherical deposits of calcium carbonate, like the spherulite crystals that Ehrenberg had first mistakenly believed to be related to coccoliths. Harting called these structures calcospherites (Fig. 2.49a). The striated calcite might grow into banded columns laced with 'fine fibres' (Fig. 2.49b), or plates of fused polyhedra which 'sometimes attain a considerable size, and are more or less curved', resembling the shells of marine gastropods (Fig. 2.49c). These structures were not purely mineral, Harting noted, but, like the shells of organisms, they also contained organic matter. He saw 'warty' and branched forms that reminded him of sponge spicules, although his drawings look rather like corals (Fig. 2.49d). In any event, none of these structures looked much like the compact, faceted forms of conventional crystals. It is hard to know exactly what processes were going on in Harting's vessels to produce this

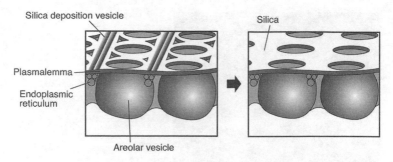

Silica deposition vesicle

Silica

Plasmalemma

Endoplasmic
reticulum

Areolar vesicle

FIG. 2.48 The formation of exoskeletons of diatoms and radiolarians is a highly orchestrated process. A froth of areolar vesicles is attached to the organism's outer membrane wall (the plasmalemma), and a scaffolding of tubular vesicles is constructed in the gaps between the vesicles. The tubular vesicles secrete silica, which forms a geometric mesh around the froth. (After Mann and Ozin, 1996.)

gallery of forms, but it seems likely that the slow diffusion of his reagents through a gel-like organic matrix resembled some of the pattern-forming processes discussed in my later chapters.

Harting hoped that his studies might pave the way to what he called a 'synthetic morphology' analogous to the synthetic chemistry by means of which chemists could mimic and reconstitute the molecules of life. D'Arcy Thompson was fascinated by the work, commenting that Harting's calcospherites 'closely resemble the little calcareous bodies' in the tissues of worms, while their fusion into polygonal plates 'closely resemble the early stages of calcification in a molluscan, or still more in a crustacean shell'. But Harting was ahead of his time: not only did he lack the means to work out exactly what his mixtures contained but there was no framework for explaining the crystallization processes that they supported. 'Synthetic morphology' did not begin to flourish until the late twentieth century, when researchers began to find controllable and reliable ways of making patterned minerals using the organizing forces of chemistry.

One of the key discoveries was made by scientists at the Mobil Corporation's chemicals laboratories in Princeton. Mobil is of course interested in oil, and one of the most powerful ways of converting the

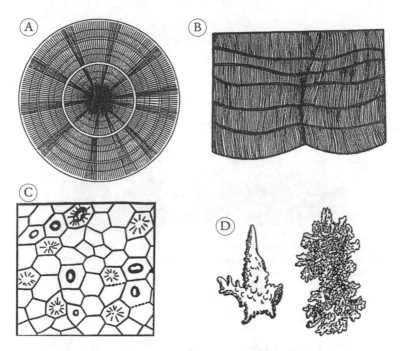

FIG. 2.49 The drawings of Pieter Harting bear witness to the extraordinary products of his experiments on 'artificial biomineralization' in the nineteenth century. He saw patterned plates like those of coccolithophores (a), fibrous bands (b), polyhedral plates (c), and 'warty growths' reminiscent of sponge spicules or coral (d). (After Harting, 1872.)

petrochemicals in oil to useful fuels and organic compounds uses catalysts called zeolites, which are minerals whose crystal structure consists of silicon, aluminium, and oxygen atoms linked into a network of tiny channels. These pores are about as wide as the hydrocarbon molecules in oil, and so zeolites can act as molecular sieves, their holes big enough to let through some molecules but not others. In the 1960s, Mobil researchers pioneered the synthesis of artificial zeolites by crystallizing inorganic solids from a mixture containing surfactant molecules. But in the early 1990s the company's Princeton team found that they could make silica molecular sieves with ordered, hexagonal arrays of pores that were much wider than those in zeolites, using relatively high concentrations of surfactants (Fig. 2.50). This

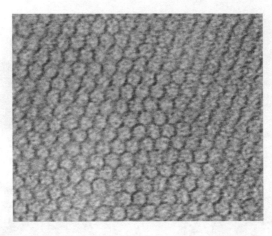

FIG. 2.50    A mixture of surfactants and silicate ions cooperate in forming a patterned solid material in which silica walls are cast around an ordered array of tubular surfactant aggregates. The material has a honeycomb lattice of microscopic pores. (Photo: Charles Kresge, Mobil Research Laboratories, Princeton.)

microscopic mineral honeycomb was generated by the spontaneous self-organizing behaviour of surfactants. The Mobil researchers concluded that these molecules were clustering into cylindrical micelles, and that the columns then packed together in a hexagonal array. Silica was precipitated around this organic mould. Other ordered structures of surfactants and polymers have been used subsequently as templates for patterning silica and other solids—for example, layered materials may be from the lamellar phase of surfactants, or bicontinuous mineral networks from periodic minimal surfaces such as the gyroid phase. In other words, the soft patterns of these organic materials can be 'fossilized' into rigid frameworks, just as radiolarians mineralize the patterns of their soft tissues.

While experimenting with synthetic techniques related to those developed at Mobil, the chemist Geoffrey Ozin and his co-workers at the University of Toronto found in 1995 that they could make minerals patterned in a quite bewildering array of forms (Fig. 2.51). These patterns, produced in aluminium phosphate, show features ranging in size from a

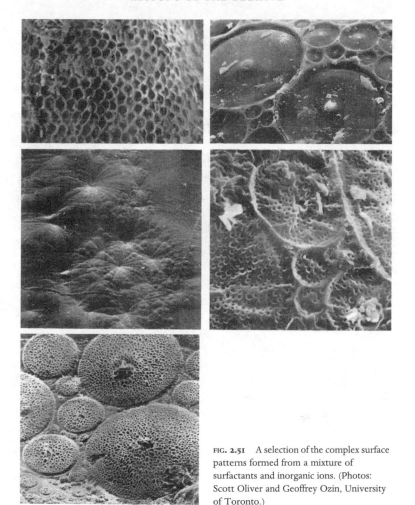

FIG. 2.51 A selection of the complex surface patterns formed from a mixture of surfactants and inorganic ions. (Photos: Scott Oliver and Geoffrey Ozin, University of Toronto.)

few millionths of a millimetre to little less than a millimetre—a hierarchy of scales echoing that found in some diatom shells. The researchers think that what they are seeing here are the imprints of a foam of organic vesicles, just like the areolar vesicles that act as templates for diatom shells.

So behind all this rich artistry, there are probably some rather simple governing forces that dictate the patterning—chief among them the self-organization of organic 'bubbles' supposed by D'Arcy Thompson.

It is a long way from intricate micro-organisms fossilized in chalk to the high-tech materials of the petrochemicals or telecommunications industries. But both of them benefit from the near-geometric regularity of the spontaneous ordering tendencies of bubbles and membranes. Haeckel's dream of a universal drive towards ordered complexity in nature was just that—a dream. Yet his inspirational drawings, for all their elaborations and exaggerations, are no fiction. They illustrate that there truly is a form of artistry, even creativity, at work in nature.

# 3

# MAKING WAVES

*Stripes in a Test Tube*

D'arcy Thompson enjoyed Pieter Harting's experiments with mashed-up oysters, but he was no chemist. (Neither for that matter was Harting, for mashed oyster is the kind of chemical reagent that only a zoologist would choose.) Indeed, *On Growth and Form* begins with something of a lament that chemistry still falls short of what Thompson considered the ultimate goal of the natural sciences: to acquire a firm mathematical basis.

This might help to account for the curious lack of chemistry in Thompson's book. But it is hard to fault him for that, since there seemed at the time to be little prospect of finding pattern in chemistry. Compared with the way that purely physical, mechanical forces can, through conflict and compromise, imprint nature with remarkable order and regularity, chemistry seems in contrast to be the great homogenizer. Pour two chemical solutions together and they mix perfectly, and the result is a bland, uniform distribution of molecules.

But not always. Thompson noted that the striations of Harting's 'calcareous concretions' seem akin to the bands or rings that may appear when certain chemicals are mixed within a jelly-like medium such as gelatin (Fig. 3.1). These structures are known as Liesegang's rings, after the German chemist Raphael Eduard Liesegang who discovered them in 1896. Thompson asserted that a similar process may account for the

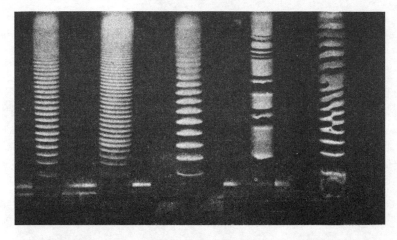

FIG. 3.1   Liesegang bands, as shown in D'Arcy Thompson's *On Growth and Form*.

banded mineral formations of agate and onyx, and for the layers that may
sometimes be found in ice deposits. 'For a discussion of the *raison d'être* of
this phenomenon, the student will consult the textbooks of . . . chemistry',
he added with a metaphorical wave of his hand—somewhat optimistic-
ally, for the effect was scarcely common knowledge and, in any case, even
if the student had found an explanation that way, it would have been
wrong. It follows, Thompson airily claimed, from 'the influence on
crystallization of the presence of foreign bodies or "impurities" '.

I do not believe D'Arcy Thompson was as uninterested in chemistry as
this discussion seems to imply. It was just that, in the early twentieth
century, there seemed to be no reason to imagine that the laws of
chemistry might contain fruitful prescriptions for patterning—and, as we
shall see, some good reasons for thinking that they could not. Yet it turns
out that, so far as the morphology of living organisms is concerned,
chemical patterning processes are far more central to our story than are,
say, the physical forces such as surface tension that shape the exoskeletons
of marine micro-organisms.

In one sense, chemistry *has* to be at the root of living form, since all life
stems ultimately from the interactions of molecules. The structure of
everything from a bacterium to an elephant must derive from a

fundamentally chemical process. But the blend of molecular genetics and evolution—the micro and the macro of living form—known as neo-darwinism does not generally invoke any *spontaneous* chemical patterning in the sense that I am using here. Until quite recently, molecular biology was envisaged largely as a system under perfect genetic control: cells were programmed at the outset to follow a developmental course that was governed dictatorially by their genes. In this picture, it was rather tempting to regard the genome as an instruction book that indicated where every cell and membrane was to be placed, and included all the assembly directions for the molecular machines that would put them there.

Yet the business of building an organism as complex as, say, a sea squirt or a fruit fly (let alone you) is now seen to be at the same time more complex and more simple than that. It seems that nature does rely to a considerable degree on the propensity of both physical and chemical systems to display spontaneous order, structure, and regularity. For chemistry is indeed a wonderful source of spontaneous patterns.

## WHY CHANGE?

Let me start by explaining why chemical systems cannot form patterns. Yes, this argument is perfectly wrong (Liesegang's rings alone demonstrate that), but I hope you will see that it is a pretty good argument all the same, and explains why the idea was resisted, in the face of the evidence, for many years.

Put simply, a mixture of chemical reagents in a beaker should become uniform throughout, attaining a state devoid of pattern. In this state there is no difference in composition from place to place. This is what we see when we let a drop of ink fall into a beaker of water and disperse. Of course, this dispersal does not happen instantly: it takes some time for the ink to spread. But it is obvious all along where the process is heading: the ink becomes ever more diffuse, and the mixing never proceeds in the opposite direction to reconstitute the original droplet. Once the transient period of 'equilibration' is over, the beaker has reached its equilibrium state, which is uniform and featureless. At equilibrium, the state of a system is stable and unchanging.

The concept of equilibrium applies not only to chemical processes but to any process of change (indeed, the spreading of an ink drop is not really a chemical process at all, but merely physical mixing). If we pick up the glass and then put it down again, we may set the liquid sloshing; but over time this movement subsides and the water surface becomes smooth and flat. A boulder rolls down a hillside into the valley, and reaches equilibrium at the bottom. Perhaps it rolls a little up the opposite side first, and then back and forth in ever-decreasing arcs: this is all just equilibration, and it will pass. At equilibrium all things are in their proper place, as Aristotle might have said, and there is no motive force to drive further change.

That processes in the world around us seem to move towards equilibrium is so familiar and so intuitive—raindrops fall to earth, rivers flow downhill, ink mixes with water—that we rarely feel the need to ask why. Or we can be placated with a tautology, which is why for hundreds of years philosophers were happy to agree with Aristotle that these things happen because it is in the nature of the elements that they should.*
I began to give a more sophisticated answer in the previous chapter, when I suggested that the equilibrium shape of a soap film is that which minimizes the surface energy. The tendency of a system to find the state of lowest energy seems at first to be a persuasive principle, and it is often invoked to account for why, say, objects thrown into the air fall back again: because of gravity, their energy—specifically, their gravitational potential energy—increases the further they are from the ground.

But, considered a little more carefully, this explanation will not do, since energy is never 'lost' in the universe; it is only transformed from one sort to another. This, in fact, is the very first law of the branch of science that describes processes of change, called thermodynamics. (The term means 'movement of heat', for the discipline arose in the nineteenth century from considerations of the relation between heat and movement or power in engines.) According to the first law of thermodynamics,

---

*In fact Aristotle's explanation invoked a hierarchy of elements, each of which has a natural place in the cosmos. Earth and water belong in the centre, so rain falls downwards, while fire rises towards the heavens. Air sits in between.

energy is 'conserved': the total energy of the universe always remains the same.*

Thus, energy is not lost when a boulder rolls downhill. The boulder's potential energy is transformed into kinetic energy—the energy of movement—as well as into some acoustic energy (the noise of its rolling) and heat (due to friction with the ground and the air). When the boulder comes to rest, the deficit in potential energy is then accounted for more or less entirely by heat. So if such processes are not truly driven by a tendency to minimize energy, what *is* causing the change to happen?

The answer is entropy. This concept is often defined, rather nebulously, as the amount of disorder in a system. Speaking more technically, as the Austrian physicist Ludwig Boltzmann did when he defined entropy formally in 1872, it is related to the number of *microstates* available to a system. All that really means is that to determine the entropy of a particular arrangement—a particular *state*—of a collection of particles such as atoms or molecules, you have to add up the number of equivalent ways of arranging the particles in that way. There are generally many more ways of making disorderly arrangements than orderly ones. Think, for example, of half a dozen apples in a crate (Fig. 3.2). An ordered arrangement might consist of placing the apples in rows. There are in this case 720 equivalent ways of doing this (we can work that out by numbering the apples and enumerating all the different permutations). A disordered arrangement consists of putting the apples anywhere, without any obvious geometric regularity. There are innumerable ways of doing

*I don't think I have ever succeeded in writing about thermodynamics without a surfeit of footnotes, for just about any simple statement in this connection needs clarification if it is not to be misleading, or downright wrong. In this instance I want to point out that the first law is typically used in reference not to the whole universe but, rather more cryptically, to a 'closed system'. This is a system from which no energy can escape, and into which none can enter from outside. You might object that this definition begs the question of whether energy is conserved or not—if you let none in or out, how could it do otherwise than stay the same? But it is not obvious that the total energy of a closed system may not spontaneously decrease, just as the population of a country can rise or fall even if no one crosses its borders. The first law, however, says that this cannot happen: the energy of a closed system remains constant. No part of the universe constitutes by itself a perfectly closed system—there is always some leakage—but we can often make this leakage insignificant, for example by insulating a sealed vessel, and thereby create good approximations to closed systems, to which we can apply the first law.

FIG. 3.2 There are many more ways of making disorderly, random arrangements of apples in a crate (*a, b*) than making orderly one as in *c* (or, in general, any other such).

that! Of course, there are other arrangements than strict rows that might be judged orderly; but you see the idea.

The second law of thermodynamics, which determines the direction of all spontaneous change (in other words, everything that happens) in the universe, says that closed systems always evolve towards states of higher entropy. So the entropy of the universe is always increasing. This tends to make things disperse and mix, rather than segregate and separate. There are many more ways of arranging the particles in an ink droplet more or less evenly throughout a beaker of water than there are of confining them all within the droplet (simply because there is then more space available to put them in). The physical process that drives this mixing is the random motion of ink particles, caused by the buffeting they receive as they collide with water molecules that are jiggling about because they are warm. These meandering trajectories make the ink particles slowly disperse, a process called diffusion. It makes intuitive sense that these random, diffusive paths of ink particles will lead to complete mixing, but in fact there is no absolute guarantee of this: precisely because the trajectories are random, there is a chance that they might all conspire to carry the ink particles back into a reconstituted droplet. But the chance of this happening is utterly negligible compared with the likelihood that diffusion will result in an even dispersion of ink. And this is precisely what the second law says: evolution towards the high-entropy, uniformly mixed state happens not because there is some cosmic imperative that demands it, but because this is an overwhelmingly more likely outcome than the reverse. Diffusion of particles in solution, it seems, promotes disorder, uniformity, high entropy, and the achievement of thermodynamic equilibrium.*

I have already given several examples of systems in which order, rather than declining as the second law seems to demand, actually increases: a regular pattern forms where there was none before. And there are many more such processes in the following pages. How is that possible, without breaking the (thermodynamic) law? To answer this we must distinguish two possibilities. One is the kind of process exemplified by the unmixing of block co-polymers to form ordered domains (pages 94–95), in which the resulting pattern is an *equilibrium* structure. There is no great mystery about this: the fact is that, by segregating into separate domains and thus surrounding themselves with molecules that are like themselves, the polymer chains may lower their energy. The lost energy escapes as heat, which warms up the surroundings. Then the loss in entropy from ordering the polymer chains is more than compensated by the gain in entropy from the heat released. So order can be bought in an equilibrium state at the cost of increasing the entropy of the surroundings. The same is true of crystals forming out of frozen liquids. The atoms in the crystals are ordered, whereas in the liquid they are disordered. But as the liquid freezes, it releases energy (called the latent heat) which warms the surroundings, enabling it to 'afford' the order of the crystalline state.

*There is another way of thinking about the second law that does not invoke this picture of probabilities of microstates, but instead says that it describes the tendency of *energy* to spread out. Concentrated energy becomes dispersed energy. There is a huge amount of energy concentrated into a lump of coal, and even more in a lump of dynamite. Add a flame, or some other appropriate trigger, and the energy spreads out. By the same token, heat energy always flows from hot to cold. This is the second law at work. The ultimate end result is that heat, like ink in water, gets homogenized, dispersed uniformly. This is what led the physicist Rudolf Clausius in the mid-nineteenth century to forecast the inevitable winding-down of the universe, which will end in a 'heat death' when there are no longer any concentrations of energy that can be used to drive useful processes—a kind of tepid Armageddon.

The key fact here is that, one way or another, an equilibrium state exists for any system. And the second law of thermodynamics dictates that this is the state for which the total entropy of the universe is maximized. Moreover, the second law seems to insist that spontaneous change will always carry the system towards this equilibrium state. This change might take a very long time—iron can take decades to turn into rust—but the direction is always clear. Thus, the second law appears to define an arrow of time: change happens in the direction that takes the universe from a low-entropy to a high-entropy state.

And now we come to the second of the two possible pattern-forming processes that I alluded to above. These are ones in which the direction of change does *not* lead us towards an equilibrium state corresponding to the state of highest entropy. Such processes, indeed, sometimes appear to be leading in the opposite direction. This looks at first glance as though it violates the second law of thermodynamics, the very principle that I have just asserted to be universal. That is exactly what scientists in the mid-twentieth century thought too, and so they were most disapproving when Boris Pavlovitch Belousov claimed to have evidence to the contrary.

## OUT OF BALANCE

Belousov, a Russian biochemist, was not looking for controversy. By all accounts he would have been glad of a quiet life; but in the 1950s he discovered something that he could not ignore, no matter how heretical it seemed.

He was interested in the metabolic process called glycolysis, by which enzymes break down glucose and capture the energy that this chemical reaction releases. Belousov devised a cocktail of chemical ingredients that was supposed to represent an artificial analogue of glycolysis, and he mixed them together. But the reaction that followed did not seem to settle down into an equilibrium state. The mixture was initially clear, and it turned yellow as the reaction proceeded. But having done so, it then turned clear again, and then back to yellow and so on, pulsing at regular intervals over and over again.

Chemical reactions, like all other processes of change, have a 'downhill' direction: that which leads to an increase in total entropy. When the reaction proceeds, the second law seems to insist that it must do so in this direction: the reaction eventually reaches the equilibrium state in which the total entropy change has been maximized.* But Belousov seemed to be suggesting that his reaction had no preferred direction: first it went one way, then the other. It was as though he was claiming to observe an ink drop that dispersed and then reformed, time and time again.

That sounded absurd, and so Belousov found himself unable to publish his findings in any reputable journal. Everyone decided that they were obviously due to his experimental incompetence. In the end, he sneaked the results into an obscure volume of conference proceedings on a completely different topic. Outside the Soviet Union, Belousov's 'oscillating reaction' remained unknown.

The irony was that Belousov's discovery was not new, and neither did it lack an explanation. In 1910 the Austrian-American ecologist and mathematician Alfred Lotka described how in theory a chemical reaction might undergo just this kind of oscillation, switching its direction back and forth. In his original model, the oscillations were 'damped'—like the decay of a ringing bell, they gradually die out and the system settles into a steady state. But ten years later Lotka showed how such oscillations might be sustained indefinitely.

Lotka was not one to defer to the imperatives of thermodynamics. 'The two fundamental laws of thermodynamics', he wrote, 'are, of course, insufficient to determine the course of events in a physical system. They tell us that certain things cannot happen, but they do not tell us what does happen.' How could he make such a claim? Lotka had the insight, obvious now but remarkable at the time, that living systems (and not just living systems) are unlike a flask of chemicals mixed together and left to react.

---

*At the level of individual molecules, chemical reactions *do* run in both directions. If the reaction involves molecule A joining to molecule B, then, were we to be able to follow the process through a microscope that gave us a molecular-scale picture, we would see both A and B combining *and* the composite AB molecules falling apart, even at equilibrium. But at equilibrium the rates of these two processes are equal, so that the average amounts of A, B and AB stay the same. Depending on the magnitude of the overall entropy change, this equilibrium state may contain a greater or lesser proportion of AB. This is an illustration of the fact that the molecular world is always dynamic, its molecules always in motion. The stasis of equilibrium is a reflection of the unchanging averages that emerge from this microscopic dynamism.

They are, in contrast, constantly acquiring energy from their surroundings: plants soak up sunlight to conduct photosynthesis, and organisms ranging from bacteria to humans gobble up energy-rich matter such as plants in order to drive their metabolic biochemical processes. '[In] systems receiving a steady supply of available energy (such as the earth illuminated by the sun)', said Lotka, 'and evolving, not towards a true equilibrium, but (probably) toward a stationary state, the laws of thermodynamics are no longer sufficient to determine the end state.' In other words, a constant flux of energy can prevent equilibrium from being reached. It is significant that Lotka says this applies not only to living systems but to our planet as a whole, bathed in the energy of the sun. We will find these prescient words, written in 1922, echoing throughout the rest of this book.

D'Arcy Thompson mentions Lotka's work in the revised edition of *On Growth and Form*, but he did not appreciate its full implications. It crops up not in any discussion of chemistry or biochemistry but in Thompson's description of animal population dynamics. For this was Lotka's principal focus; his scheme for an oscillating chemical reaction was proposed primarily as an analogue of the way populations of animals interact, as though they are no more than molecules: 'When the beast of prey A sights its quarry B, the latter may be said to enter the field of influence of A, and, in that sense, to collide with A', Lotka explained.

Lotka formulated a series of equations that showed how oscillating population sizes might arise in these 'colliding' predators and prey. His work was extended in the 1930s by the Italian biologist Vito Volterra, who showed how the scheme could be used to understand fluctuations in fish populations. I shall return in Chapter 5 to this description of the dynamics of ecosystems.

When he presented his theory of persistent oscillations in 1920, Lotka referred in passing to the fact that 'in chemical reactions rhythmic effects have been observed experimentally'. He gave no details, and frankly I do not know to whose experiments he was referring. But such oscillations were certainly reported the following year by the chemist William Bray at the University of California at Berkeley. Bray was certainly no bungling experimenter: he pioneered the study of the rates of chemical reactions, and his student Henry Taube became a Nobel Laureate.

Yet when Bray found that a chemical reaction between hydrogen peroxide and iodate seems to deliver its products, oxygen and iodine, in pulses, he got a reception as cool as Belousov's 30 years later. Bray even cited Lotka's work as evidence that such a thing was possible, but to no avail.

It is a testament to the respect and awe with which scientists regard the second law of thermodynamics that the reality of oscillating reactions such as those proposed by Lotka took half a century to become established. During the 1960s, the biochemist Anatoly Zhabotinsky, then a graduate student at Moscow State University, came across Belousov's buried findings and decided (for graduate students tend not to have their preconceptions set in stone) to take them seriously. Zhabotinsky discovered a mixture of chemical compounds that produced a colour change far more striking than the rather insipid transformations of Belousov's solutions: the oscillations went from blue to red. This concoction, in which the organic compound malonic acid reacts with bromate, with metal atoms added to catalyse the reaction, is now known as the Belousov–Zhabotinsky (BZ) reaction. If the BZ reagents are combined and mixed well, the solution switches from red to blue and back every few minutes.

Something as dramatic as this was hard to ignore, and thanks to Zhabotinsky's insistent advocacy chemical oscillations became accepted during the 1970s as something that really happens. In 1980 Belousov and Zhabotinsky, together with their colleagues Albert Zaikin, Valentin Krinsky, and Genrikh Ivanitsky, were awarded the prestigious Lenin Prize by the Soviet government. That was not quite the happy ending it seems, for Belousov died ten years earlier, while his discovery had not yet gained wide recognition.

## THE CHEMICAL SEE-SAW

How, then, does the BZ reaction elude the second law? Well, it doesn't, and neither does any known process, whether physical, chemical or biological. If we leave the BZ reaction blinking away, we find that the oscillations do not last for ever. Eventually (it can take hours) the mixture will settle into a steady, unchanging state of equilibrium—and this is indeed a state in which the entropy of the flask and its surroundings has increased. It is simply that the mixture takes a long and circuitous route to reach this destination.

Lotka was on the right track, but he did not quite express the matter correctly. It is true that thermodynamic laws do not necessarily tell us 'what does happen', but this is because they speak only about end points. They tell us what is the equilibrium state of a system, whether that is a flask of chemicals or a planet. But they do not say anything about how that state is reached, about how the change unfolds. As chemists know, thermodynamics alone is of limited value in comprehending chemical change. It is not of much practical use, for example, to know that two reagents can be combined to make a thermodynamically stable product, if that process takes a thousand years to occur. The rate and manner with which chemical reactions take place are the subjects of the discipline known as chemical kinetics. Understanding the BZ reaction is thus a question of unravelling the kinetics of the process.

The oscillations of the BZ reaction are not unavoidably doomed to fade away. They can be sustained indefinitely if the pot is constantly replenished with the raw ingredients of the reaction, and if the end product (primarily, malonic acid to which bromine atoms have been added) is removed. Chemists have devised vessels that sustain a constant throughflow of materials, which are well mixed in a reaction chamber: they call them continuous stirred-tank reactors, or CSTRs. It is not hard to see that we too—each one of us, each living human being—are fundamentally CSTRs, filled with a chemical mixture of staggering complexity that is compartmentalized into highly intricate channels and chambers. Our genes are not what sustains our lives: they would be helpless if our internal 'stirred tanks' were not constantly replenished with raw materials (sugars, amino acids, vitamins, salts, oxygen, water) and flushed free of wastes (not just the obvious ones but also compounds such as the carbon dioxide we exhale).

This continual flux of materials through a CSTR prevents the chemical system from reaching equilibrium. You have never in your life experienced true personal equilibrium, for if you had, your life would be over. Equilibrium is deathly; nothing happens there. Equilibrium for the universe means Clausius's heat death, a cosmos rendered totally homogeneous. Scientists care about equilibrium states, but to the rest of our world they are anathema. All life exists out of equilibrium, and as Lotka observed, this is ultimately made possible on Earth by the continual flux of energy from the sun.

This is what makes the planet itself come to life. It is why water circulates between the sky and the seas, why the winds blow, why plants grow and why the biosphere sustains the atmosphere in a state of extreme disequilibrium. You cannot keep an atmosphere full of a compound as reactive as oxygen for millions of years without some non-equilibrium process to replenish it: it would otherwise react with rocks and get bound up in the solid earth. This is why scientists searching for life on other planets believe that an atmosphere with a high oxygen content is a likely signature. No need to search for cities, roads, radio broadcasts: the atmosphere alone can give the game away. As indicated in Chapter 1, some researchers have suggested that the best way to look for life on Mars is not to sift through the soil for bugs but to analyse the chemical environment for signs that it is not at equilibrium.

What Lotka showed is that a chemical system out of equilibrium can develop a kind of pattern: in this case, a pattern in time, a regular oscillation in composition. We will see shortly that non-equilibrium chemical reactions have a far more general pattern-forming potential than this. But first we need to look at where the oscillations come from. What is it that makes Lotka's hypothetical mixture, and the very real chemical cocktail devised by Belousov and Zhabotinsky, so indecisive about which direction to take?

## BLOW UP

Like a great many chemical reactions, the BZ reaction depends on the process known as *catalysis*. A catalyst is a substance that speeds up the rate of a chemical reaction without being changed itself. There is a nice (if messy) physical analogy for this: if you pour a fizzy drink into a glass and then add a few grains of sand, the fizzing becomes more vigorous. The sand grains act as a kind of catalyst for bubble formation. They provide sites where bubbles can form more easily, which is to say, they lower the energy needed to initiate a bubble. A chemical catalyst likewise lowers the energy needed to initiate the formation of the reaction products from the initial reagents. Most industrial chemical reactions use catalysts, since they would otherwise run too slowly to be economically viable. And almost all biochemical reactions in the body are assisted by natural catalysts, the proteins called enzymes.

What makes the BZ reaction unusual is that it makes its own catalyst. This means that one of the product molecules acts as a catalyst to speed up the formation of more product: it is self-catalytic, or *autocatalytic*. This is an example of a positive feedback process, which is self-amplifying. Left to its own devices, autocatalysis makes the process go ever faster. A nuclear explosion involves this sort of positive feedback, as do most chemical explosions. Autocatalysis is prone to literally blowing up out of control.

But how can autocatalysis lead to oscillations, rather than simply to a runaway process? There needs to be some way of checking the propensity for blowing up. Let me illustrate what Lotka had in mind here by using his own context of predator and prey populations. Let's say that there is a population of rabbits and a population of foxes that prey on them. Rabbits are notoriously autocatalytic: rabbits lead to more rabbits, or in other words, rabbits catalyse their own production. Given unlimited grass, a rabbit population will multiply exponentially.

Now enter the foxes. They eat rabbits; and well-fed foxes multiply too—in other words, foxes lead to more foxes, but only so long as there are rabbits around to sustain them. This is also an autocatalytic process, but it depends on the presence of rabbits. Yet foxes, no matter how well fed, die off from time to time: there is a steady rate of attrition of the fox population.* So we can write three 'equations' to describe how the populations of predator and prey change over time:

1. Rabbits and grass lead to more rabbits
2. Rabbits and foxes lead to more foxes
3. Foxes lead to some dead foxes

Each of these processes happens at a characteristic rate.

Suppose now that we start off the ecosystem with a few rabbits and a few foxes. The rabbits multiply quickly, and their numbers rise. This

---

*The same is true of rabbits, of course, but because rabbits multiply so fast, we do not need to incorporate this into the picture.

means that the foxes have an abundant source of prey. In line with step 2, the fox population also starts to rise. The trouble is that the foxes don't know when to stop: they gorge on rabbits, and decimate the rabbit population. And then they find that there is suddenly no food, and so step 2, which relies on the presence of rabbits, can no longer take place. According to step 3, the foxes already present gradually die off. Now, this *can* lead to a situation in which the ecosystem drives itself extinct: the foxes eliminate all the rabbits, and then they all starve and die. This certainly can happen in the wild. But suppose the foxes don't quite manage to kill off all the rabbits—some of them elude their predators. When there is a big fox population but few rabbits, there is not enough food for all the foxes and so their population crashes. This gives the few remaining rabbits some respite, and their numbers begin to rise again. We are back to the situation we started with: lots of rabbits and few foxes. Then step 2 kicks in again and the fox population rises while the rabbits decline. Within a certain range of value of the relative rates of the three steps in the process, the system undergoes regular oscillations in the numbers of both foxes and rabbits, with the fox population peaking almost perfectly out of step with the rabbits (Fig. 3.3).

This is precisely the scheme that Lotka used to show how sustained chemical oscillations might occur. He replaced rabbits and foxes with chemical compounds; for example, suppose that a reaction of compound

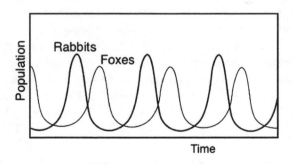

FIG. 3.3 Alfred Lotka's mathematical treatment of predator–prey populations gives rise to population sizes that oscillate out of step with one another.

G[rass] with compound R[abbits] leads to more of R (an autocatalytic step). We can write this as

   1. G + R → more R

In the second step of the process, R reacts with compound F[oxes] to make more F (another autocatalytic step):

   2. R + F → more F

Finally, compound F decays spontaneously into another compound, D[ead foxes]:

   3. F → D

This is the same scheme as that for the case of rabbits and foxes, and will likewise produce oscillations in the concentrations of reagents R and F. Now suppose that R is coloured red and F is blue—then we have a process that changes colour back and forth, just like the BZ reaction.

The crucial point to appreciate is that the oscillations are sustained by a steady influx of materials or energy. In the case of the rabbits and foxes, the rabbits can go on multiplying only so long as there is grass. This keeps growing because of the influx of solar energy, water, and so forth. And the system never gets clogged up with the carcasses of dead foxes, because these decay and return to the soil. The ecosystem is like a CSTR in which materials (compound G) are constantly added and waste (compound D) is constantly removed. This is what prevents the system from reaching a static, unchanging equilibrium. Instead, it reaches a dynamic *steady state*—there is no stasis, but the same thing keeps repeating.

The BZ reaction is not exactly like Lotka's scheme: it is a great deal more complex, involving at least 30 different chemical compounds and many steps. When Western scientists learned about the reaction at an international conference in Prague in 1968 that brought them together with the Soviets, the chemists Richard Field, Endre Körös, and Richard Noyes at the University of Oregon set about deducing its mechanism. By 1972 they had devised a somewhat simplified scheme which accounted for the oscillations. Two years later, Field and Noyes pared this model down to an even simpler one, called the Oregonator (the 'Oregon oscillator'), in which there are just five steps involving six chemical compounds, most of them containing bromine and oxygen. Just one of these steps is autocata-lytic, and rather oddly, none of them involves malonic acid (recall that the 'end result' of the BZ reaction is to add bromine atoms to this organic acid). The bromination of malonic acid does not feature in the oscillatory part of the process—it is simply a side reaction induced by the 'output' of the Oregonator, a chemical species called hypobromite.

The key feature of the Oregonator is that it has two branches—two distinct sets of reactions—one of which induces the red colour and one the blue. The system switches back and forth between these branches as they rise to dominance and then exhaust themselves, rather as Lotka's ecosys-tem switches between a dominance of rabbits and of foxes.

The simplest way to record the BZ oscillations is to merely keep note of the colour changes: red–blue–red–blue and so on. A more precise way is to measure the concentrations of the various compounds that wax and wane, like the numbers of rabbits and foxes (as in Fig. 3.3). But there is yet another way, too, which brings out a more profound feature of what we might call the 'mathematical form' of this process. Rather than plotting a graph of the concentration of a chemical species over time, we could plot how its concentration changes in relation to that of one of the other oscillating species. It is clear that the fox population is high when the rabbit population is low and vice versa, but in fact there are also times when both are medium-sized and either rising or falling. If we plotted the number of foxes against the number of rabbits in Fig. 3.3, we would get something like Fig. 3.4a. Each cycle of oscillation corresponds to a single

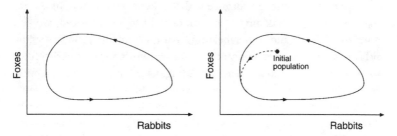

FIG. 3.4 Oscillations of two interacting populations can be depicted as limit cycles (a). If the population sizes start from a position off the limit cycle, they are soon drawn onto it (b). In this way, the limit cycle is said to act as an attractor.

circuit of this loop, which is called a limit cycle. We would get a similar limit cycle by plotting the concentration of two of the oscillating compounds in the BZ reaction. If we start the reaction off with concentrations of these reagents that lie off the limit cycle, the concentrations will evolve along a trajectory that takes them onto the loop (Fig. 3.4b). For this reason, the limit cycle is said to be an *attractor* of the system.

## GOING PLACES

The oscillating BZ reaction can thus be considered to display a periodic pattern of sorts, but it is a pattern in *time* (which is why the reaction has been dubbed a 'time crystal'). The oscillatory process can, however, also generate patterns in space. So far, I have talked only about a well stirred medium, which has a uniform composition throughout the reaction vessel at any instant. But if the reaction is carried out without stirring, small variations in the concentrations of the reagents are likely to arise from place to place. This is true of any chemical reaction, and normally does not lead to anything remarkable.

For an autocatalytic process like the BZ reaction, however, small variations can make a big difference. The positive feedback can have the effect of blowing up small differences into big ones. In particular, it means that one region of the BZ mixture can get flipped onto a different branch from that of the surrounding regions. The blue branch can appear in a sea of red. Then we have a mixture in which the colour changes from place to place.

What took chemists by surprise is that these colour variations do not take the form of a random patchwork of red and blue. Instead, we find complex, orderly, and rather beautiful patterns emerging. In a shallow dish of the BZ mixture, concentric rings or twisting spirals radiate outwards from a central source like ripples (Fig. 3.5 and Plate 2). The chemical oscillations give rise to moving *chemical waves*.

These patterns were first described by the German scientist Heinrich Busse in 1969, although it was not until the following year that Zhabotinsky and Zaikin correctly identified them as chemical waves—travelling 'fronts' of chemical change. It is not hard to see how a fluctuation in the relative concentrations of the reacting species might shift the BZ reaction from one

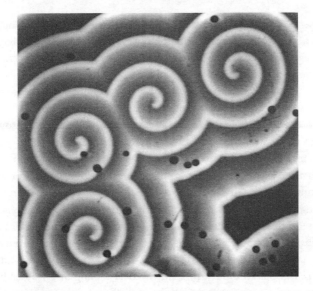

FIG. 3.5 Chemical waves in the Belousov–Zhabotin-sky reaction. Here I show examples of spiral waves, but in general both spirals and concentric target patterns may appear (see Plate 2). (Photo: Stefan Müller, University of Magdeburg.)

branch to the other; but why does this disturbance then radiate outwards as an organized wave?

Imagine that a small region of the solution has been tipped from the red to the blue branch. Because of the autocatalysis, this blue region expands from its origin as the molecules diffuse into the 'red' solution and induce tipping there. The rate of diffusion is equal in all directions, and so this expanding wavefront is circular. The notion that autocatalytic reactions can generate propagating chemical waves in fact predates any recognition of oscillating reactions. In 1900 the German physical chemist Wilhelm Ostwald showed that, when he pricked the dark surface of oxidized iron immersed in acid with a zinc needle, this triggered an electrochemical process that changed the colour of the surface coating, which propagated away from the point of contact at high speeds. From the 1920s some researchers studied this system as a simple analogue of nerve impulses, which are also electrochemical waves. Meanwhile, in 1906 Robert Luther, the director of the Physical Chemistry Laboratory in Leipzig, announced the discovery of chemical waves in autocatalytic reactions to an audience of German chemists in Dresden. Some were sceptical, until Luther demonstrated the process before their eyes, projecting an image of a chemical wave on to a screen.

Luther pointed out that the waves depend on competition between the autocatalytic process and the diffusion of molecules. Autocatalysis can rapidly exhaust the available resources: if the grass doesn't grow fast enough in the first step of Lotka's scheme, the rabbit population might crash through lack of food even without the foxes preying on them. Luther said that the same can happen in the corresponding chemical process—except that here the 'grass', which we called reagent G, can diffuse into the growing rabbit colony from the surroundings to replenish the supply. Because of this delicate balance between the rate of reaction and the rate of diffusion, chemical waves are said to be a *reaction–diffusion* phenomenon. After Luther's pioneering studies, the theory of reaction–diffusion systems was placed on a firm mathematical footing in the 1930s by the Russian mathematician Andrei Kolmogoroff and the English geneticist Ronald Fisher. Like Lotka and Volterra, Fisher was interested in these systems because of their relevance to population dynamics: he was investigating the rate at which an advantageous gene would spread through a population. Clearly, biologists shared none of the hesitation that chemists displayed in assimilating these ideas about complex dynamics and pattern formation.*

Autocatalysis and the reaction–diffusion process account for how an expanding reaction front can become a *wave*, where the concentration of a chemical species rises and then falls again as the wave passes. Within the wavefront, the autocatalytic process takes hold and then rapidly exhausts

---

*This might seem strange, given the aversion to mathematics and abstract theory that biologists have often shown, from D'Arcy Thompson's time to our own. But ecologists, population biologists and to some extent neuroscientists have always been unusual in this regard, embracing mathematics in a way that biologists working at the level of cells and molecules have not. Lotka's 1924 book *Elements of Physical Biology* was the first exposition on what later became known as mathematical biology (its 1956 reprint used this term instead), while Fisher made important contributions to pure statistics as well as population genetics. We will encounter several other examples in later chapters, and in the other books in this series, of biologists of this persuasion blazing a trail into the world of pattern formation well in advance of physical scientists.

itself. All the while, the region just in front of the wave remains 'ripe' for colonization by the diffusing autocatalytic species—and so the wave progresses outwards. But in the BZ medium this happens not once but repeatedly: once the source of the first wave is established, it goes on discharging subsequent waves at regular intervals, giving rise to a well defined wavelength (the distance between successive wavefronts).

This, as you have probably guessed already, is because the BZ reaction is not simply autocatalytic but oscillatory. The wave source region is like a tiny flask of BZ mixture, blinking blue and red—but this 'flask' has no boundaries, and so the transformation propagates. You might wonder why all of the mixture does not behave the same way, so that it becomes a chaos of wave sources. The reason is that, once the first wavefront has passed through a part of the mixture, that region is 'enslaved' to the 'pacemaker' at the wave's origin. Behind the wavefront, the medium has been 'exhausted' by the passing wave and cannot switch branches again until the oscillatory cycle has played itself out—and it is at precisely this point that the next wave arrives. Each wave source expands its territory until it encounters a wavefront from another source. When the waves collide, they annihilate one another. This is because one wave cannot excite the region behind the other wavefront, which is in the 'exhausted' phase of the cycle. So the collisions of waves create fixed, stationary boundaries between territories commanded by each pacemaker.

An unstirred BZ mixture can thus be considered to consist of three types of region. One is at the wavefront itself, where autocatalysis induces branch switching and a change of colour. Here the medium is said to be 'excited', rather like the electrical surge of a nerve impulse. In front of the wave, the medium is ripe for excitation: it is in a 'receptive' state. And behind the wavefront, the medium is exhausted or 'refractory', resistant to further excitation until the cycle has run its course.

A medium that has the potential to adopt these three states is called 'excitable', and is liable to experience the circular, periodic travelling waves characteristic of the BZ mixture.* The ingredients for a model of

---

*In experiments on the BZ system, the reagents are generally infused in a layer of gel. This slows down the rates of diffusion, and makes the chemical waves more stable and less sensitive to disturbances, so that they have a smoother, regular shape.

an excitable medium are rather basic and generic—they say nothing about the particular chemical reagents involved. In fact, they do not even specify a *chemical* process at all. Researchers have studied computer models of a sort of general-purpose excitable medium which is represented as a checkerboard lattice of little compartments or cells, each of which interacts with those around it. The 'rules' of the model are:

1. Each cell can be in either an excited, a receptive, or a refractory state.

2. Excited cells become refractory after a certain length of time, and must stay that way for a fixed period until returning to the receptive state.

3. Receptive cells are transformed to the excited state if a certain proportion of their neighbours are excited.

This sort of model, in which an array of cells adopt specific states contingent on the states of their neighbours, is called a *cellular automaton*, reflecting the fact that the cells' behaviour is conditioned by an automatic, knee-jerk response to those around it. It is an extraordinarily versatile way to model systems that consist of many interacting components, and I shall draw on cellular-automaton schemes repeatedly in the following pages. The cellular automaton for the BZ reaction captures the essential characteristics of an excitable medium—but there is nothing in the prescription that gives any hint of the kinds of patterns that might arise. Yet when this model is run on a computer, it produces just the kind of target and spiral patterns seen in the real BZ medium (Fig. 3.6).

This shows that the wave pattern has nothing to do with any features of bromate or malonic acid or any of the other ingredients. It is to be expected for any system that has the characteristics of an excitable medium. The patterns are *universal*.

The account I have given so far explains why we should expect concentric target waves, but it is not clear why spiral waves are generated as well. The spirals are actually 'mutant targets', created by a disruption of the circular wavefront. Such perturbations can happen by accident, for example if there is some impurity such as a dust particle in

FIG. 3.6 The wave patterns of the BZ reaction are mimicked in a mathematical model called a cellular automaton that takes no account of any chemical details, but simply represents the mixture as a grid of cells that can be 'excited' by receiving stimuli from their neighbours. (Image: Mario Markus and Benno Hess, Max Planck Institute for Molecular Physiology, Dortmund.)

FIG. 3.7 In three dimensions the BZ reaction produces rather complex patterns, the simplest of which is this spiralling scroll wave. (Image: Arthur Winfree, University of Arizona.)

the reaction medium; or they can be induced on purpose, for instance by blowing air on to the wavefront through a narrow pipe. At the break in the circular wave, the ends curl up and become sources of spirals.

If the BZ mixture is infused not into a thin layer of gel but into a thick slab, then the chemical waves may propagate in three dimensions. The patterns are then more complex. A spiral wave, for example, becomes a three-dimensional form called a scroll wave (Fig. 3.7). Cross-sections of scroll waves look like concentric target waves in one plane, and like pairs of counter-rotating spiral waves in another. These patterns were first seen in BZ mixtures in the 1970s.

# GAS RINGS

Other chemical reactions that share with the BZ process both autocatalytic steps and a competition between different reaction pathways can also generate oscillations and wave patterns. These include some combustion and corrosion reactions, and also biochemical processes: oscillations can indeed be seen in the glucose-splitting metabolic reactions that Belousov first set out to study.

In the early 1990s, the German chemist Gerhard Ertl and his co-workers at the Fritz Haber Institute in Berlin discovered an example of chemical waves in a reaction of considerable technological interest, the conversion of carbon monoxide and oxygen to carbon dioxide, catalysed by platinum metal. This is essentially the process that takes place in an automobile's catalytic converter, where poisonous carbon monoxide is removed from the exhaust fumes. Ertl's group had already found hints of such behaviour several years earlier, when they saw that the formation of carbon dioxide could be oscillatory, coming in pulses. When they used a new kind of microscope to look at the progress of the reaction on a platinum surface, they found target and spiral patterns in the distribution of carbon monoxide (CO) molecules and oxygen atoms attached there (Fig. 3.8).

This is curious, because there is nothing obviously autocatalytic about the transformation: oxygen atoms and CO molecules simply combine on the metal surface to make carbon dioxide. But the researchers found that when CO molecules stick to platinum, they change the arrangement of metal atoms at the surface. And, critically, this new arrangement is more effective at catalysing the transformation of CO to carbon dioxide. So in effect CO assists its own transformation. Irregularities on the metal surface may then spark off chemical waves.

One difference between an oscillating reaction happening in a solution and on a metal surface is that the solution looks the same in all directions, whereas on a crystalline metal all directions are not equivalent: the regular stacking of metal atoms creates a kind of checkerboard grid with particular symmetry properties. That is why the target and spiral patterns in the oxidation of CO on platinum are elliptical rather than circular: the speed of the chemical waves differs in different directions. This effect of surface *anisotropy* (non-equivalence of directions) is even more visible in the

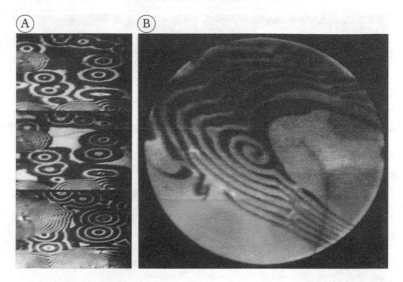

FIG. 3.8 Target (a) and spiral (b) waves in the reaction of carbon monoxide and oxygen on a platinum surface. These images are several tenths of a millimetre across. (Photos: Gerhard Ertl, Fritz Haber Institute, Berlin.)

chemical wave patterns generated by the reaction of nitric oxide and hydrogen on the surface of rhodium metal (Fig. 3.9), where the square symmetry of the metal surface imposes itself on the wavefronts. We will see other examples below of how the changes may be rung on pattern formation by the imposition of an underlying symmetry on a universal class of patterns.

Oscillations in the burning of carbon monoxide to make carbon dioxide can happen even in the absence of a platinum catalyst—that is, when the pure gases are mixed and ignited in a CSTR. This creates regular surges in temperature, which can leap up by many tens of degrees and then fall again in a matter of seconds. Autocatalysis here comes from complex reactions between the gas molecules that involve reactive intermediates called free radicals.

This kind of process may also create spatial patterns. As early as 1892, two scientists named A. Smithells and H. Ingel described a gas flame that separated into bright patches, like flower petals, which rotated slowly around the flame's axis. The flame was hotter in the bright regions than in

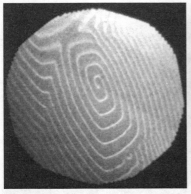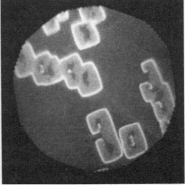

FIG. 3.9 The effect of the anisotropy of the metal surface—the symmetry of the packing of surface atoms, which makes some directions different from others—can imprint itself on the form of these chemical waves, as shown here in the square appearance of spiral waves in the oscillating reaction of nitric oxide and hydrogen on rhodium. (Photo: Ronald Imbihl, University of Hannover.)

the darker boundaries between them. In the 1950s George Markstein of the Cornell Aeronautical Laboratory studied these 'cellular flames', and in 1977 the Israeli scientist Gregory Sivashinsky showed that the cell patterns could be produced by a reaction–diffusion process: because of the different rates at which oxygen and hydrocarbon molecules like butane diffuse through space, the fuel can become depleted in some regions of the flame.

A reaction–diffusion process should be capable of producing ordered cell patterns, whereas those seen by Markstein and others were random and constantly growing, shrinking, and coalescing. But in 1994, researchers at the University of Houston in Texas found that indeed a cylindrical butane flame could break up into a variety of regular patterns, which became increasingly complex as the flow rate of the gas was increased (Fig. 3.10). These patterns could undergo strange shifts: in those with concentric rings of cells, for example, the rings could rotate independently in hopping motions, like gear wheels shifting position. Or the ordered cells might break up into a disorderly arrangement of randomly shaped cells, only to reform moments later. Other researchers have seen spiral patterns in flames that look like those of the BZ reaction, reinforcing the idea that a reaction–diffusion mechanism is responsible.

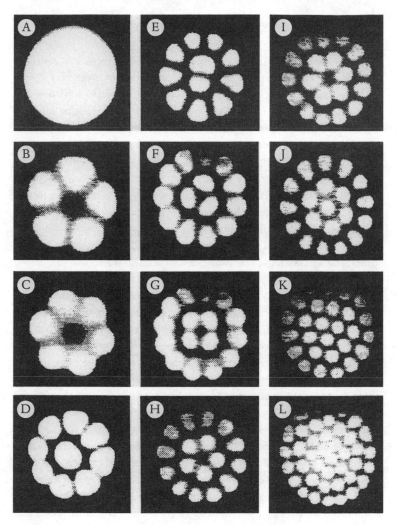

FIG. 3.10 Cellular patterns in a shallow cylindrical flame. The temperature is lower in the darker regions. (These are not truly dark to the eye, but are evident in the video recordings.) The cellular flames adopt ordered states, which increase in complexity as the rate of gas flow in the flame increases (from a to l). (Photos: Michael Gorman, University of Houston.)

# RINGING THE CHANGES

Raphael Liesegang's banded structures perhaps now seem less puzzling. He found them while investigating the precipitation of silver salts from a solution of silver nitrate in gelatin, a process closely related to that which occurs in photographic emulsions. Liesegang's father, Friedhelm Wilhelm Eduard, and grandfather were both early experimenters in photography and magic-lantern projections, and in 1854 Friedhelm Liesegang founded the Photographic Institute in Elberfeld, Germany; this became the company Liesegang Technology, which still manufactures projectors and display screens today. Raphael Liesegang was by all accounts a remarkable, not to say eccentric, character with interests as diverse as D'Arcy Thompson's. He speculated about the possibility of television and, in addition to his work on photography he studied bacteriology, chromosomes, plant physiology, neurology, anaesthesia, and silicosis.

Liesegang found that a drop of silver nitrate placed on a flat layer of gel infused with potassium chromate solution produces dark deposits of insoluble silver chromate in concentric rings. One might expect that as the silver salt diffuses through the gel, it would simply generate a circular dark patch of silver chromate that becomes more diffuse at the edges as the silver is depleted. But, instead, the precipitate appears in a series of rings like the growth rings of a tree trunk. In a cylindrical tube filled with gel, these rings become a series of bands as the silver nitrate descends through the column, as shown in Fig. 3.1. Several other precipitation reactions behave similarly when carried out in a gel (Fig. 3.11).

This chemical patterning captured the imagination of several leading scientists of the fin-de-siècle Victorian era, including Lord Rayleigh, J. J. Thomson (the discoverer of the electron) and Wilhelm Ostwald. Some commentators suggested that the bands and stripes might represent a simplified version of the markings one can see on zebras, tigers and butterflies—an idea that led one critic to complain in 1931 that 'enthusiasm has been carried beyond the bounds of prudence'. True, this was speculation far beyond what the evidence warranted; but if it was imprudent, it was nevertheless close to the mark.

Given that the gel slows down the rate of molecular diffusion, one might suppose that Liesegang's bands are an example of a reaction–

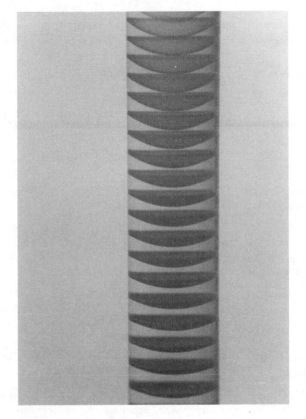

FIG. 3.11  Liesegang bands produced in a column of gelatine laced with a cobalt salt. They are caused by oscillatory precipitation of cobalt hydroxide, as a reagent diffuses down the column. (Photo: Rabih Sultan, American University of Beirut.)

diffusion process—and this is almost certainly the case. But the precise details are still not fully understood today. One of the first explanations was proposed by Ostwald, an expert on crystal precipitation and growth, just a year after Liesegang reported the experiments. According to this theory, the precipitating particles must reach a critical size before their continued growth is guaranteed: below this size (the critical nucleus), a particle is as likely to fall apart again as it is to keep growing. Ostwald imagined the silver nitrate descending in a wavefront down the column, and reacting as it goes to produce silver chromate. This insoluble compound will not precipitate as small crystals until they grow to the critical size. But this growth is slowed by the gel, which lowers the rate at which the ions diffuse towards a crystal nucleus. So the gel gets over-concentrated (supersaturated) in silver chromate, until finally a threshold is crossed and the concentration is high enough to trigger crystal formation (nucleation) everywhere. Then almost all the silver chromate is flushed out in a pulse, producing a band, and the concentration remaining in solution plummets far below the threshold. It takes some time for this concentration to build up again by diffusion of silver nitrate, by which time the reaction front has moved on. So there is a cycle of supersaturation, nucleation, precipitation, and depletion that dumps a train of bands in the wake of the advancing front. It's a little like the person who, paid daily, saves up his cash for a weekend binge that leaves him starting each week freshly impoverished.

Ostwald's theory is most probably correct in essence, though it has remained a challenge to explain all of the phenomena associated with the Liesegang patterning process. For example, in 1923 K. Jablczynski investigated the 'rhythm' of the bands, showing that if one measures the position of successive bands down the column (relative, say, to the first of them), then the ratio in the positions of two consecutive bands approaches a constant value. Jablczynski argued that Ostwald's theory could explain this 'law'; but when his theory was further refined by German-American chemist Stephen Prager in 1956 it predicted that the bands should be infinitely narrow, whereas in fact their thickness tends to increase down the sequence. In 1994 Bastein Chopard of the University of Geneva and his co-workers devised a way to describe the process using a cellular-automaton model which takes into account the microscopic processes that control the diffusion, nucleation, and precipitation of the reacting species. This model generates a series of widening bands that obey Jablczynski's spacing law (Fig. 3.12). But band spacings in real experiments do not always follow this law—in fact, there seems to be no general law that applies to all Liesegang-type processes. So there is still work to be done on this most venerable of artificial chemical patterns.

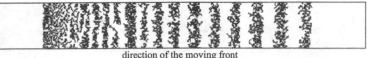

direction of the moving front

FIG. 3.12 Liesegang bands generated in a computer model of the precipitation–diffusion process. (Image: Bastien Chopard, University of Geneva.)

## ROCK ART

Was D'Arcy Thompson right to suggest (as in fact Liesegang did himself) that Liesegang rings are related to the banded patterns of minerals such as agate? It sounds plausible, for these minerals are formed by precipitation from a supersaturated solution of metal salts, which in turn is created when warm water flows through fissures in cooling basaltic lava and flushes out the metals. The solubility of the salts is lower at lower temperatures, and so the minerals precipitate as the salt-rich water cools.

There is now good reason to believe that many banded minerals do arise from cyclic precipitation out of mineral-laden water. The bands found in some iron oxide minerals, the wood-grain texture of cherts, the striations of zebrastone, and the concentric rings of agates have all been explained this way. For example, geologists Peter Heaney at Princeton University and Andrew Davis at the University of Chicago have shown that Liesegang-style precipitation–diffusion cycles can explain the iridescence of iris agates. The pearly appearance of these minerals is due to a periodic banded structure on a scale too small to see with the naked eye. The widths of the bands are about the same as the wavelength of visible light, which means that the banded structure scatters light. The wavelengths scattered most strongly, and therefore the apparent colour of the banded material, depend on the angle from which it is viewed. The same effect creates iridescence in opal (which is made up of regularly packed arrays of tiny silica spheres) and on the wingscales of butterflies, which are covered in microscopic ridges. The effect was explained in the late nineteenth century by Lord Rayleigh, and D'Arcy Thompson himself hinted that Liesegang-like periodic banding might be responsible for mineral iridescence.

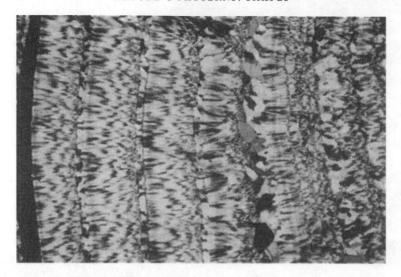

FIG. 3.13 Liesegang banding in iris agate. The image is about 2.5 mm across, and the vertical bands are due to layers of quartz alternating with thicker layers of chalcedony. The horizontal striations have a different origin, caused by the fibrous structure of the mineral. This banding gives the mineral an iridescent appearance. Agate typically has banded structures on several different scales, most familiarly the concentric patterns visible to the naked eye (see Plate 3). (Photo: Peter Heaney, Princeton University.)

In the case of iris agates, the bands record differences in the crystal structure of the mineral (Fig. 3.13): quartz alternates with a form of chalcedony that is peppered with flaws in the crystalline ordering of atoms. These defects are caused by the formation of chain-like molecules from the silicate ions that constitute both materials—something that happens when the concentration of ions is high. The poorly crystalline chalcedony precipitates rapidly, which depletes the silicate solution close to the surface of the growing mineral, producing conditions that favour precipitation of quartz. But the quartz precipitates slowly, allowing a build-up of silicate ions until conditions are ripe for making chalcedony again. In effect, just as in Ostwald's explanation of Liesegang's rings there is a rapid process that overshoots and induces a 'dead time' before it can start again, generating oscillations that are imprinted on the growing mineral.

FIG. 3.14   These samples of jasper look like stones carved by ancient craftspeople. But the patterns are formed naturally during crystallization of the mineral from a gel. (Photo: Manuel Velarde, Universidad Complutense, Madrid.)

Agates are often banded on many length scales: the iridescence comes from bands a few micrometres (thousandths of a millimetre) wide, but there are also bands at scales of several hundred micrometres and a few millimetres (Plate 3). Apparently there is a *hierarchy* of oscillatory patterning processes at play here. We saw the same range of patterning scales, for quite different reasons, in diatom shells, and we will see below that this kind of hierarchy is fairly common in natural patterns.

A different sort of mineral patterning can arise from the interaction of reaction– diffusion processes of solidification with the flow of a gel-like solution from which the rocks form. The pieces of jasper shown in Fig. 3.14 could easily be mistaken for primitive carvings, or even for the fossils of some strange form of life. But these patterns are purely inorganic. The German physicist Hartmut Linde and his wife Gudrun recently found these striking objects in the eastern deserts of Egypt, and in collaboration with physicist Manuel Velarde in Madrid they deduced how these natural sculptures were made. The researchers suggest that the jasper, another kind of chalcedony, forms as silicate ions diffuse through a gel-like fluid, link together, and precipitate. The ridged formations are the result of the flow properties of the paste-like chalky sediments in which the jasper forms: pressure builds up in the gel by diffusion of silicate ions until it is able to push back the surrounding sediment, whereupon the rock solidifies and the pressure is released—only to rise subsequently as silicate accumulates again. It is not unlike the regular sticking and slipping of a violin bow as it moves across the strings, which produces acoustic oscillations.

## WILD AT HEART

We saw above that cellular flames can display a range of different patterns, switching from one to another as the rate of gas flow is increased. These jumps are not gradual but abrupt: first there are five cells, say, and then suddenly a sixth has appeared. This illustrates another important general feature of self-organized patterns: the patterned system may turn out to have several alternatives to choose from, and it 'changes its mind' suddenly. What determines this choice? That is one of the most profound questions of the field, and I shall return to it more than once. For now, let's just say that switches or *transitions* between patterned states may often be induced by increasing the force that drives the system away from its equilibrium state.

This is a complex statement, so I'll clarify. Let's go back to the BZ reaction in a continuous stirred-tank reactor, oscillating periodically between a uniform red and blue. Here, remember, it is the flow of reagents through the tank that ensures the chemical mixture does not reach equilibrium—as indicated above, if the flow is turned off, the solution eventually settles down to an unchanging equilibrium state. So then, what happens if we increase the factor that sustains the reaction away from its equilibrium point, namely the flow rate?

The answer is that the oscillations, while appearing to continue as before, actually acquire a subtly different character. Suppose we are monitoring them by measuring the concentrations of various chemical species that rise and fall with each cycle—bromide will do. We find that, once the flow rate exceeds some critical threshold, the pulses of bromide acquire a double rhythm, rising to the same level only every other peak (Fig. 3.15a). The change is revealed even more clearly if we look at the limit cycle (page 124)—let's say, the circulating variation in the concentration of bromide plotted against another oscillating ingredient such as bromite.

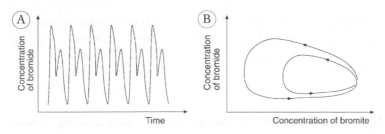

FIG. 3.15   As the flow rate of a BZ mixture through a continuous stirred-tank reactor is increased, the oscillations double up, a phenomenon called period doubling (a). This is reflected by the appearance of a second loop in the limit cycle (b).

Instead of one loop, the cycle now has two (Fig. 3.15b). One complete cycle of the system corresponds to a circuit of both loops.

This behaviour is called period-doubling: the periodic oscillations have a double pulse. The transition from a single (period-1) to a double pulse (period-2) is called a bifurcation. That is simply a fancy way of saying 'splitting in two', which is just what has happened to the limit cycle. Period-doubling bifurcations are extremely common in systems that undergo non-equilibrium oscillations, as they are driven further from equilibrium. Such a doubling-up of the BZ reaction was first observed in the 1980s by the French physicist J. C. Roux and his co-workers.

In fact, even the simple back-and-forth oscillations of the period-1 state are achieved via a bifurcation in the BZ mixture. As we have seen, the long-term state of this mixture in a closed vessel is not oscillatory but steady and unchanging: the oscillations gradually die away. And that will also happen in a CTSR if the flow rate is too slow; only above a critical threshold do the oscillations persist indefinitely. This kind of abrupt transition from a constant state to an oscillatory one was identified by the German mathematician Eberhard Hopf in 1942, long before it became clear that chemical oscillators exist. It is now called a Hopf bifurcation. Below the Hopf bifurcation, the state of the system can be described by a single point on a graph of its parameters, like the concentrations of chemical species in Fig. 3.15b. If these concentrations are perturbed, they will gradually return to their values at this 'fixed point'. But above the bifurcation, the system starts to 'wobble': the fixed point becomes a loop, as though the system can no longer find its way back to that point but keeps overshooting and undershooting, orbiting it for ever.

Increasing the flow rate of a BZ mixture in a CSTR therefore switches it first to a period-1 state, and then to period-2. And it doesn't stop there. Boost the flow further and there is another period-doubling bifurcation, this time creating four concentration peaks in each full cycle, and four lobes on the limit cycle. At a still higher flow rate, another period-doubling transition takes us to eight peaks per cycle. The 'distance' (that is, the amount by which the flow rate must increase) between these jumps decreases for each successive bifurcation, and as the jumps get closer and closer together, the periodic pattern gets more and more complex. Eventually there comes a point at which all periodicity is lost: the system still 'oscillates' in the sense that the chemical concentrations rise and fall, but there is no predictable pattern to it (Fig. 3.16). The system has then become *chaotic*.

This sequence of transitions—from a steady state to oscillations of increasing complexity, and ultimately to chaos, via period-doubling bifurcations, as the out-of-equilibrium driving force is increased—is a very common kind of behaviour in pattern-forming systems. We will come across it again.

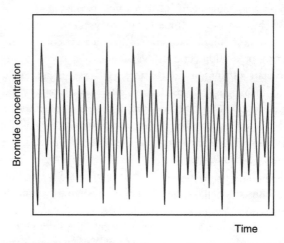

FIG. 3.16 At high flow rates, the oscillations of the BZ reaction become apparently random—the system becomes chaotic.

## RHYTHMS OF LIFE

We have now some reason to suspect that chemical waves and oscillations are more than a quirk of the arcane mixtures devised by Belousov and Zhabotinsky. Nonetheless, one might imagine they require rather specialized conditions and are unlikely to be particularly relevant to everyday phenomena.

We shall see in the following chapters that this could not be further from the truth. Reaction–diffusion processes represent one of nature's most widespread and versatile schemes for producing patterns, and the concepts that underpin them are not restricted to chemical systems. I can demonstrate that very simply. The fact that you are reading these lines shows that there is at least one analogue of a chemical wave pulsing away in your body at this very moment, as it was yesterday and the day before that, and as it will until the day you die. Indeed, though I am sorry to say it, the fact is that it will quite probably be the breakdown of this wave-forming process that will be the cause of your demise: your heart will lose its ability to sustain regular muscular contractions. Your heartbeat is, loosely speaking, a chemical wave, and your heart a kind of slab of Belousov–Zhabotinsky gel. And the pattern it makes is your lifeline.

The beat of a heart is a remarkable feat of synchrony. With each pulse, a wave of electrical activity that begins in a pacemaker region called the sinoatrial node surges through your cardiac tissue. At the front of this travelling wave, tiny molecular channels open up in the walls of your heart cells and electrically charged ions pass through, moving from one side of the membrane to the other. This alters the voltage across the cell walls and induces contraction of the cardiac muscle. In order for the heart to be effective as a pump, these contractions must be highly orchestrated: it would be no good if each individual cell decided for itself when to alter its voltage, so that the heart merely flutters freely like a frightened bird. The coordination, expressed as a travelling wave, is attained because cardiac tissue is an excitable medium, just like a BZ mixture. Once the wavefront has passed, the cells enter a refractory phase during which they reset their membrane voltage by redistributing the ions inside and out.

This analogy between the heart and the BZ reaction has provided one of the major motivations for studying the latter: scientists are not merely interested in making pretty chemical patterns but hope that these might, as a model system amenable to laboratory study, help us to understand some of the vagaries of the heart.

These are all too familiar. Heart attacks are the leading cause of death in industrialized nations, and the majority of them stem from a pathological condition called cardiac arrhythmia. Crudely speaking, this is the state of a heart that has lost its beat. In the final stages of arrhythmia, called ventricular fibrillation, the coordinated electrical changes that produce a regular pumping motion have vanished and the heart is reduced to a futile, quivering mass. Despite its name, however, arrhyhmia does not initially lead to a total loss of rhythm: rather, a new rhythmic activity appears called tachycardia, in which the regular beats—about one every second in adult humans—give way to rapid pulsations about five times faster (Fig. 3.17). If unchecked, arrhythmia gives way to uncoordinated, weak fibrillation, which usually ends in sudden cardiac arrest. The Scottish pioneer of cardiac electrophysiology John MacWilliam, who first showed the connection between ventricular fibrillation and heart attacks, described this fateful turn of events in 1888 in rather poignant terms: 'The cardiac pump is thrown out of gear, and the last of its vital energy is dissipated in a violent and prolonged turmoil of fruitless activity in the ventricular wall.'

It appears that this fatal condition is associated with the conversion of regular travelling waves in the heart to spiral waves. You can see from Plate 2 that spiral waves in an excitable medium tend to have shorter wavelengths than target waves (and thus a more rapid pulse, if the waves move at the same speed). As a result, spiral waves, once created, come to dominate and displace target waves because they 'jump in' and excite the medium more quickly. A clue to the role of spiral waves in ventricular fibrillation is that, when cardiac arrhythmia is initiated, the frequency of the heart's oscillations increases.

These lethal spiral waves have now been seen directly in beating hearts. By keeping fragments of rabbit heart tissue active in a solution of salts and nutrients infused with oxygen, in 1972 researchers at the University of

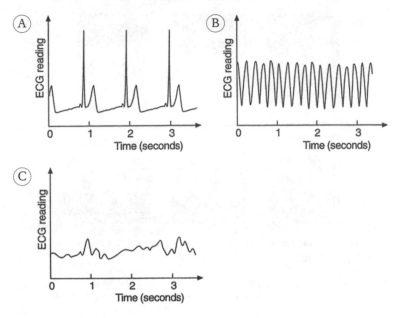

FIG. 3.17 The onset of ventricular fibrillation, which causes heart attacks, can be seen in the changing oscillations of heartbeats, as revealed by electrocardiograms (ECGs). A healthy heart beats regularly about once a second (*a*). This can switch to faster, but still regular, pulsations in the condition called cardiac arrhythmia (*b*). This behaviour may then dissolve into feeble, uncoordinated oscillations in ventricular fibrillation (*c*), which is quickly fatal if not arrested.

Amsterdam saw a wave of electrical activity spinning like a turbine blade at ten revolutions per second. In 1995, Richard Gray and his co-workers at the State University of New York Health Science Center in Syracuse showed that a single rotating spiral wave could give rise to ventricular fibrillation in whole rabbit and sheep hearts sustained in a culture medium. They revealed the patterns of electrical activity propagating through the hearts by adding to the artificial 'blood' supply a dye that emitted fluorescent light with a brightness proportional to the local voltage. The researchers saw spiral waves of electrical activity with centres that meandered over the heart surface (Fig. 3.18*a*). The electrocardiograms associated with this behaviour showed the uncoordinated oscillations characteristic of fibrillation. The spiral waves are much clearer in

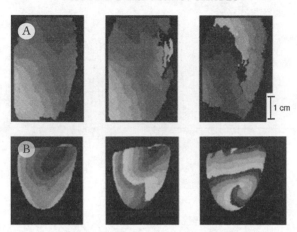

FIG. 3.18 Spiral waves in the heart may be discerned by using voltage-sensitive fluorescent dyes, which here produce a grey scale proportional to the local voltage (a). The spiral pattern can be seen rather more clearly in a computer model of the heart. (b). These spiral waves are characteristic of arrhythmia and ventricular fibrillation. (Images: Richard Gray, State University of New York Health Science Center.)

computer calculations conducted by Gray and colleagues using equations that were known to describe the basic properties of heart activity (Fig. 3.18b).

Spiral waves, then, are bad news. But what causes them? In view of what we have seen so far, it is understandable that heart tissue—an excitable medium—has spiral waves (and consequently arrhythmia and fibrillation) as one of its possible patterned states. But normal, healthy hearts will beat reliably for billions of cycles without falling prey to spirals. Their regular travelling waves correspond to a stable *attractor* state, like the limit cycle of the BZ reaction. We can think of attractors as wide valleys in a hill range: just as a ball kicked around anywhere in the valley will tend always to roll back down to the valley floor, so the heart can be subjected to all manner of jolts without risk to its healthy pulsating state. But the spiral waves of ventricular fibrillation are another attractor— another valley just over the brow of the hills. Kick the ball too hard, and it will cross the ridge and roll down into the new attractor. In the same way, severe shocks to the heart, such as are encountered by a person struck by lightning or undergoing heart surgery, can trigger this transition to a life-threatening state. In that event, another severe shock can jolt the organ back to its healthy rhythm, and that is precisely how arrhythmia is generally treated by implantable defibrillators, devices designed to give the heart a kick if it shows signs of losing its rhythm.

But diseased or abnormal hearts are typically more prone to switching into the dangerous arrhythmic state. It is as if their hill slopes have become eroded or breached, so that the spiral-wave attractor can be reached more easily. One possible cause, identified by Leon Glass and his co-workers at McGill University in Montreal, is poor communication between cells. The researchers found in 2002 that cultured layers of embryonic chick cardiac cells produce travelling waves that have a tendency to break up into spirals when they are exposed to the compound heptanol, which reduces the electrical interactions between neighbouring cells and thus makes it harder for them to synchronize their activity. Alternatively, spiral waves can be triggered when normal travelling waves are interrupted, just as we saw above in the BZ mixture. Small regions of 'inert' tissue in the heart, such as damage caused by blood clots, can act as initiating points for spirals. This damaged tissue does not necessarily present an obvious danger by impeding the flow of blood in the heart, or suppressing the regular contractions of the rest of the organ, but it can trip up the heart's patterns of electrical activity, with potentially fatal results.

The heart, like any multi-celled organ or organism, is a joint effort: all the cells have to work together. When real cells behave like those in a cellular automaton model, responding to the state of their neighbours, then they may act like an excitable medium, meaning that travelling waves can impose on them an oscillatory synchrony. Brain cells, like heart cells, communicate electrically, which is why waves of electrical activity are also characteristic of the brain: brain waves are not just a figure of speech.

It now seems clear that the communities of cells that make up a pumping heart can organize their activity in many subtle ways. As well as the regular travelling waves of electrical activity that promote the healthy heartbeat and the spiral waves that threaten to disrupt it, cardiac tissue can exhibit other wavelike patterns that display a gradual progression from coordination to chaos. In cardiac tissue cultivated in Petri dishes, Kyoung Lee and co-workers from Korea University in Seoul have seen spiral-wave states that show the kind of period-doubling observed in the BZ mixture under an increasing throughflow rate of nutrients. In other words, these 'pieces of heart' can display a doubled or quadrupled 'heartbeat', as well as the completely irregular, chaotic pulsations characteristic of fibrillation. That is really no more than we might expect from an excitable medium of this sort; but understanding what controls these switches between oscillating states is here a matter of life and death.

## THE PULSE OF THE CROWD

Our hearts and minds, and our other organs too, rely on cellular coordination and synchrony. Some cells, however, seem to get on just fine by themselves. Bacteria, which are single-celled, are in fact the most pervasive and successful organisms on the planet. We humans do quite well in weathering the parched Gobi desert and the frozen Arctic, but it curbs our pride to known that some bacteria live amidst hot oil hundreds of feet underground, while others thrive in the ice above buried Antarctic lakes, or in superheated water around submarine volcanoes, or in the vats used to store radioactive waste.

Yet individualism has its limits, and there are times when even bacteria are better served by talking and cooperating. If food is scarce, bacteria may decide to behave rather like ants, foraging in groups. Bacteria are not, however, noted for their powers of communication. They are blind, deaf, and dumb. They make up for these limitations with what we might rather crudely call a splendid sense of smell. 'Bacterial talk' is conducted in a chemical language: each of them emits a kind of perfume called a chemo-attractant, much as animals emit pheromones to attract mates. Not only can other bacteria detect this material but they can also sense where it is coming from. They merely follow their noses, or, to put it less anthropomorphically, they detect and follow a concentration gradient. Like ink seeping into blotting paper, the chemo-attractant released by a bacterium gets more diffuse the further away it travels, and so one bacterium can find its way to another by following the upward slope in concentration, using thrashing whip-like appendages to propel itself. This chemically stimulated movement is called *chemotaxis*.

Bacteria are not the only cells to employ chemotaxis. Our own cells communicate this way too, which is how they organize themselves into complex structures such as branching neural dendrites in the brain. So does the slime mould *Dictyostelium discoideum*, an amoeba that lives in soil, feeding on bacteria. When deprived of heat or moisture *Dictyostelium* cells begin a chemically guided process of conglomeration that sees them gather into multi-celled bodies better suited to harsh conditions. This pile of identical cells decides on a division of responsibility: they 'differentiate' into different cells types, some becoming part of a bulbous 'fruiting body' supported on a long stalk, like a weird kind of mushroom. The fruiting body contains spores, which can survive without food or water in suspended animation, waiting for better days.

This is somewhat analogous to the way cell specialization and segregation into tissues takes place during the growth of an embryo of a more complex organism, for which reason *Dictyostelium* has been regarded as a simple model of the kinds of developmental processes that I shall discuss in later chapters.* But here it is the early stages of aggregation that concern us, because these turn out to involve the appearance of patterns that will doubtless now look familiar (Fig. 3.19).

Where do these targets and spirals of slime come from? In the first stages of chemotactic cooperation, some of the cells (called, for obvious reasons, pioneer cells) release pulses of a chemical compound with a complex name that can be mercifully abbreviated as cAMP.** The formation of cAMP is an enzyme-catalysed process that is autocatalytic: cAMP can, in effect, boost the activity of the enzyme that makes it. That is why the formation of the signalling molecule comes in oscillatory bursts.

As this chemo-attractant diffuses out of pioneer cells, other nearby cells follow it to its source. Crucially, once a pioneer cell has emitted a burst of cAMP, it falls 'silent' for several minutes, as if to recuperate. This is a 'refractory' period, and it means that *Dictyostelium* behaves like an excitable medium—which is why it exhibits the generic target and spiral waves. Yet here these patterns are a passing phase: in time the cells

---

*The chemotactic mechanisms of patterning in *Dictyostelium* are, however, quite different from those found in the embryogenesis of multi-celled organisms.

**For the curious and the chemically inclined, this is cyclic adenosine monophosphate.

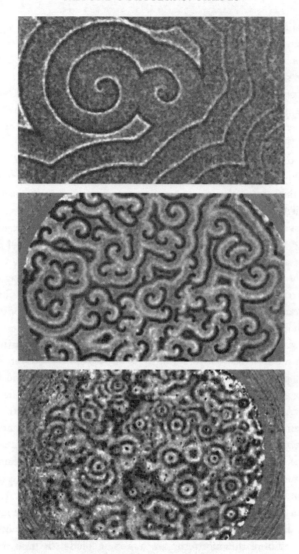

FIG. 3.19 Target and spiral patterns in colonies of the slime mould *Dictyostelium discoideum*. These patterns are generated when some cells emit periodic pulses of a chemical attractant, towards which other cells move. (Photos: Cornelis Weijer, University of Dundee.)

form trails that converge on the central source of the waves, creating cell piles that then differentiate into slimy mushroom bodies. The period of excitable behaviour is transient, merely a means to a quite different end.

Chemotactic bacteria are not limited to making targets and spirals, but have a far wider repertoire of patterns. In 1991 the biologists Elena Budrene and Howard Berg at Harvard University showed that the prosaic human gut bacterium *Escherichia coli* displays astonishing artistry. The researchers grew colonies of *E. coli* in layers of gel (which again slowed down their diffusion) under life-threatening conditions, such as too little food, too much oxygen, too little heat, or exposure to chemicals that disrupt the regular biochemical processes of the bacterial cells (poisons, you might say). They found that the cells would cluster into spots and streaks patterned as if rendered on shields or plates by the most inventive of Islamic artists (Fig. 3.20 and Plate 4). Unlike the travelling waves of *Dictyostelium*, these patterns remain stable for long periods. Budrene and Berg assumed that they are the result of chemotactic signalling between the bacterial cells, which emit a chemo-attractant called aspartate when stressed. But the sheer complexity and variety of the patterns—much richer than those of *Dictyostelium*—baffled everyone.

Yet two teams of physicists devised models that could reproduce some of the basic features of the patterns by assuming that they arise from competition between a few basic processes: cell multiplication by division, cell migration by diffusion in search of food, and cell clustering by chemotaxis once the local density of cells (and thus the local concentration of chemo-attractant) exceeds a certain threshold. A group led by Eshel Ben-Jacob at Tel Aviv University in Israel postulated that the cells move around as groups of 'walkers' that consume nutrients in the environment, and reproduce. If food becomes scarce each walker emits a chemotactic signal, and responds to the signals of others by 'climbing' up concentration gradients. At the University of California at San Diego, Herbert Levine and Lev Tsimring proposed much the same model but described it using reaction–diffusion equations, similar to those studied by Ronald Fisher in the 1930s in his work on genes and populations, rather than considering discrete 'particles' of bacterial cells. Both teams found that their models predicted that the colony, starting from an initial cluster of cells, will expand in a ring which breaks up into spots behind the advancing front

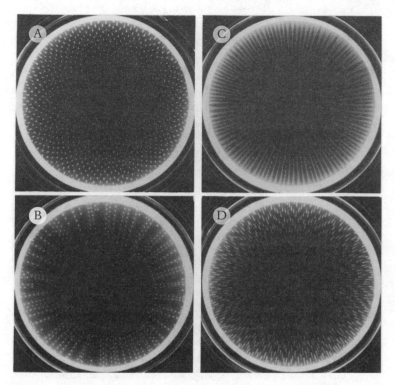

FIG. 3.20 Symmetrical patterns formed by the bacteria *Escherichia coli* in response to chemical signalling. (Photos: Elena Budrene, Harvard University.)

once the chemotactic signalling is turned on (Fig. 3.21*a*). Ben-Jacob and colleagues found that these spots would become aligned in radial streaks (Fig. 3.21*b*), as seen in some of Budrene and Berg's experiments, if they added a further element to the model: a *repulsion* between the walkers, resulting from the emission of another chemical signal that this time warns others to stay away. Do *E. coli* really emit such a chemo-repellent? That is not clear: it's a prediction that has not been verified.

Budrene and Berg found that their *E. coli* patterns do seem to be generated in a process resembling that predicted by these models. The colony advances, as predicted, in a ring (called a swarm ring) which breaks up spontaneously into cell clusters once the bacteria begin emitting the

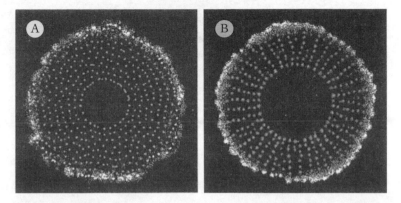

FIG. 3.21 Concentric spots (a) and radial patterns (b) of clustering bacteria can be reproduced in a computer model of cell communication and movement. (Images: Eshel Ben-Jacob, Tel Aviv University.)

chemo-attractant aspartate. Nevertheless, Budrene felt that a model that accounts for this behaviour should include nothing more than the known biological properties of the bacterial cells, rather than, say, a speculative repulsion. In collaboration with Michael Brenner and Leonid Levitov at the Massachusetts Institute of Technology, she argued that the swarm-ring expansion and break-up could be explained by a reaction–diffusion scheme in which the cells move in search of a nutrient called succinate, which they convert to the chemo-attractant aspartate. The spots arise from a positive feedback process in which slight irregularities in cell density produce local 'hotspots' of chemo-attractant, raising the density even further. This clustering is called an instability: the uniform swarm ring is bound to break up into spots once the point of instability is reached, regardless of the precise details of, for example, how strong the attractive interaction between cells is. It's a little like the Rayleigh instability that fragments a column of liquid into a string of droplets (page 60).

No one knows how varied the kaleidoscope of bacterial patterns might be. *E. coli* is by no means the only bacterium to display them: *Salmonella typhimurium*, for example, also shows a rich array of spots and rings, while we will see in Book III that *Bacillus subtilis* colonies display a quite different kind of 'universal form'. It seems likely, however, that all these patterns arise from the same basic processes characteristic of reaction–diffusion: random dispersal competes with interactions between the components that cause clustering and create instabilities. I will say it again: patterns come from a delicate balance between competing forces.

## IN THE BEGINNING

Chemical waves are with us from the start. When the eggs of all higher organisms, including humans, are fertilized by sperm, this triggers a series of waves of calcium ions that pulse over the surface of the egg every few minutes, sometimes persisting for hours. The purpose of these waves is not clear, but they are thought to be somehow priming the egg for further development into an embryo. The waves are mostly simple ripples that travel from one side of the egg to the other; but because they are generated in a reaction–diffusion process, more complex patterns are possible, such as the now-familiar spiral wave (Fig. 3.22).

Spirals are ubiquitous in nature, but that does not mean they are all formed in the same way: the whirlpools of fluid flow, for example, are quite different beasts. It is tempting to imagine that spiral galaxies (Fig. 3.23) are merely whirlpools writ large, stirred up in a rotating mélange of dust, gas, and stars. But, remarkably enough, it seems possible that these cosmic patterns are in fact also the product of a kind of reaction–diffusion process—that they are patterns related to those in the

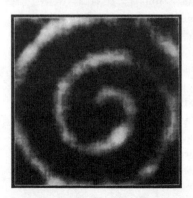

FIG. 3.22 Spiral waves of calcium travel across the surface of frog eggs when they are fertilized. The precise purpose of these waves in embryo development remains unclear. (Photo: David Clapham, Mayo Foundation, Rochester.)

FIG. 3.23 The spiral galaxy NGC 5236 in the southern sky. The structure of some spiral galaxies may be the result of a star-formation process that mimics a reaction–diffusion system. (Photo: European Southern Observatory.)

Belousov–Zhabotinsky mixture. In the 1960s Chia-Chiao Lin and Frank Shu suggested that the spiral arms do not rotate 'rigidly', so that stars are either inside or outside the arms, but, rather, they are waves of high density that sweep through the galactic material like sound waves through air—otherwise the galaxy would be pulled apart by centrifugal forces as it rotates.

But there is another possibility that might operate at least in some spiral galaxies: the bright arms are not simply waves of high density of matter but regions where new stars are forming. This then introduces positive and negative feedbacks. Dust produced in the atmospheres of existing stars feeds new material into the surrounding interstellar medium that could promote further star formation—a kind of autocatalysis for stars. But at the same time, the light emitted by stars heats up this medium and so makes it expand rather than contract into stars. Meanwhile, behind the bright regions of star formation are dark zones where stars have grown old and dim, and have depleted the interstellar medium of the material needed for more stars to form. So star formation here remains quiescent until it is replenished by fresh material. In other words, this is a galactic excitable medium. In 1996 the physicist Lee Smolin pointed out that a galaxy can indeed be regarded as a kind of reaction–diffusion system, and its cosmic pinwheels the manifestation on an immense scale of the whorls that might appear on an egg that has just been triggered into becoming a person.

Yes, here are life's universal patterns. We'll be seeing more of them.

# 4

# WRITTEN ON THE BODY

*Hiding, Warning, and Mimicking*

D'arcy Thompson was unsatisfied by answers to the 'why' of pattern and form that stopped at adaptive function. He complained, for example, about the way embryologists would insist that every aspect of a developing organism must, as he put it, 'be referred directly to their usefulness in building up the body of the future animal'. To Thompson this was little better than a 'Just So Story'.

The criterion of *usefulness* was indeed what Rudyard Kipling had in mind when he explained how the animals of Africa got their distinctive pelts:

> After another long time, what with standing half in the shade and half out of it, and what with the slippery-slidy shadows of the trees falling on them, the Giraffe grew blotchy, and the Zebra grew stripy, and the Eland and the Koodoo grew darker, with little wavy grey lines on their backs like bark on a tree trunk; and so, though you could hear them and smell them, you could very seldom see them, and then only when you knew precisely where to look.

Kipling's 'explanation' is in fact a glorious mixture of ideas. He gives a mechanism for the pattern's formation: the shadows of trees are somehow imprinted on the animal skins, I suppose by a kind of tanning. But he also suggests a reason why this pattern persisted: it was *useful* to the creatures as a

FIG. 4.1   Zebras in Kenya. (Photo: Michael and Sandra Ball.)

form of camouflage, increasing the chances of survival. A striped zebra merges with the long grass and bushes of the veldt (Fig. 4.1), while a spotted leopard is better adapted to sneaking up on its prey in a sun-dappled forest.* This all sounds resoundingly Darwinian, although it would be more accurate to view Kipling's tales as Lamarckian, describing environmentally acquired characteristics that may be inherited.

Kipling, inadvertently I am sure, thus summarizes all the various points of view about animal patterning advanced in the nineteenth and early twentieth centuries. Or almost so—for what he could not conceive of (and who can blame him?) is the essence of D'Arcy Thompson's argument: that a pattern might form itself. Kipling's markings are made either via a pre-existing template—the trees act as a mask that blocks out some sunlight, like a stencil—or by being individually put in place. (The leopard, you may recall, is finger-painted with pigmented spots by the Ethiopian with whom he hunts, so that he might not show up in the dark forest 'like a mustard-plaster on a sack of coals' as he stalks his prey.)

---

*Just how effective the zebra's stripes are as camouflage remains a matter of debate. The British zoologist Hugh Cott (who knew a lot about camouflage, enough to have advised the military on it during the Second World War) quoted one observer as saying 'in thin cover he is the most invisible of animals. The stripes of white and black so confuse him with the cover that he is absolutely unseen at the most absurd ranges.'

Kipling's explanations for the *mechanisms* behind the formation of these markings are of course delightfully fanciful. But Darwinism seemed to have nothing whatsoever to offer in their place. It accounts for how advantageous patterns and forms and structures, once established, might persist in the population. But on the issue of how they arose in the first place—the question of 'efficient cause' that has exercised philosophers ever since Aristotle—the theory remained silent, beyond implying a blind search among a palette of unspecified extent. This would, taken at face value, ask us to presume a bewildering diversity of proto-zebras far back in time, covered with markings of all shapes and sizes: hexagons, perhaps, or—who knows?—stars and stripes, or alphabetical letters. Are the patterns of biology truly limitless? We have already seen how similar patterns and forms may emerge in systems that seem to have nothing in common. Do zebras, giraffes and leopards draw on a particular repertoire of pelt designs, or was Darwin right to invoke the image of 'endless forms most beautiful'?

Neo-Darwinism now tells us how zebras inherit stripes from their (evolutionarily successfully) parents: via the transmission of genetic material, the instructions for how to build an organism. But the 'stripe' genes are not as deterministic as those that define the zebra's body shape—for a zebra's legs and ears are always in the same place as those of its parents, but its stripes are not. These genes, it seems, do not provide a paint-by-numbers prescription for where pigment must go, but merely convey a predisposition towards stripiness.

Although D'Arcy Thompson did not express it in quite this way, he was interested in exactly how a zebra acquires a tendency for stripiness, or a leopard for spottiness. You could say that he did not ask *why* zebras in general have stripes (which one can answer in Darwinian terms), but *how* a particular zebra gets its stripes. Here, however, he was pretty much stumped. The question of animal marking patterns does not even crop up in *On Growth and Form* until the book is virtually over, and then Thompson does little more than spend a couple of pages describing the different types of stripes that zebras exhibit before remarking vaguely and with an air of wishful thinking that 'like any other aspect of form, pattern is correlated with growth, and even determined by it'.

Yet Thompson set us on the right trail, for when he discussed Liesegang bands he pointed out that the German botanist Ernst Küster had believed these to be echoed in the regular pigmentation patterns of the living world: the striped leaves of plants such as Eulalia grass, the bands on feathers or a cat's skin, the designs of tropical fish. Küster, says Thompson, regarded them all as 'so many instances of "diffusion-figures" closely related to the typical Liesegang phenomenon'. Thompson does not commit himself here, for it seems even his inventive mind could not see how to progress beyond the superficial similarities. It was, however, only a decade after the publication of the revised edition of *On Growth and Form* in 1942 that another genius well ahead of his time saw how that might be done.

## FROZEN WAVES

Alan Turing was one of the few mathematicians to have genuinely changed the way we see the world. It usually takes an exceptional mathematician to find a way of applying the abstract and often rather arcane theorems of that discipline to the real world; but Turing did more or less the converse. He showed that what appear to be real-world, even prosaic questions are actually illustrations of what one can call a mathematical level of reality. Take computers. The notion of a mechanical calculating machine goes back at least to the eighteenth century, and when work began on electronic versions in the 1940s most people considered this an engineering problem, a matter of replacing cogs and levers with electronic switches made from diodes. Turing, however, had already shown in 1936, at the age of 24, that computation was in fact an abstract mathematical concept that had implications for the deep structure of mathematics and logic itself, for example in terms of whether theorems were inherently provable or not. He introduced the idea of a general-purpose digital computer—now known as a universal Turing machine—whose capabilities were not contingent on the details of how it was implemented in hardware. Turing's research in this area underpins today's information revolution, and laid the foundations of the field of artificial intelligence.

Modern computation is a central tool in the current understanding of pattern formation: we will see time and again how reliant this understanding is on models that can be simulated on a computer. But Turing's contribution to the field goes far deeper than that. His dream of making a thinking machine, an artificial brain, led him to think about brain structure and development in general, and thus about the question of how biological form arises—the problem of morphogenesis.

Turing knew what D'Arcy Thompson had said on the matter, since he read *On Growth and Form* as a schoolboy. Like Thompson, Turing could see that the matter of building a form as complex as the human body was doubtless an intricate and messy business that could not be expected to yield to clean mathematical analysis. But, like Thompson, Turing's interest was focused on what seemed to be a more tractable aspect of the issue: by what agency does the whole affair begin?

The various tissues and organs of a multi-cellular body share the same genetic material while clearly they serve different functions. The reason is that their patterns of gene activity are different: some genes that are actively read and translated into proteins in one kind of cell—liver cells, say—are 'silent' in other cell types, such as bone-making cells. These cells have become specialized, or 'differentiated', so that they show different regimes of gene *expression* (gene-to-protein conversion). Geneticists and cell biologists interested in the development of organisms commonly seek to identify these differences in gene expression and to understand how the different cohorts of protein enzymes in each cell type give the cells their various functions.

But how do cells that are initially identical give rise to daughter cells that are different? How do they know which genes to turn on or off? The problem is particularly acute if we extrapolate back to the very early embryo, a ball of identical cells. How does this homogeneous cluster turn into a mixture of different cell types, each with a specific location and developmental destiny?

That was the problem Turing confronted in 1952 in a ground-breaking paper called 'The chemical basis of morphogenesis'. Here he expressed the puzzle in terms so direct and blunt that one has to suspect a certain wryness:

An embryo in its spherical blastula stage has spherical symmetry, or if there are any deviations from perfect symmetry, they cannot be regarded as of any particular importance ... But a system which has spherical symmetry, and whose state is changing because of chemical reactions and diffusion, will remain spherically symmetrical for ever ... It certainly cannot result in an organism such as a horse, which is not spherically symmetrical.

The problem, then, was why there is apparently a spontaneous breaking of symmetry. How, without any outside disturbance, can the spherically symmetric ball of cells turn into one that is less than spherically symmetric, with its cells differentiated to follow distinct developmental paths?

We have already now seen how such a thing may happen. A well-stirred Belousov–Zhabotinsky mixture is initially uniform, yet if left to sit it develops spatial patterns of markedly different chemical composition. By a curious coincidence, Turing began to ponder the question of chemical symmetry-breaking just as Belousov was discovering his colour-changing cocktail (but before Zhabotinsky showed that it could produce spatial patterns). But Belousov, as we have seen, was working in obscurity in the Soviet Union, and Turing had no more knowledge of oscillating chemical reactions than anyone else. He had to invent the whole notion from scratch.

We can see from the quotation above that he had already glimpsed the necessary ingredients: chemical reactions and diffusion. It was Turing's paper, in fact, that established the term 'reaction–diffusion system'. He considered that, during development of an organism, genes in different cells might be switched on or off by chemical agents called morphogens that diffuse through the tissues. There could, for example, be a 'leg-evocator' morphogen that tells cells to develop into a leg. Other morphogens, Turing speculated, could influence skin pigmentation.

The pivotal point in his argument is that the production of morphogens might be autocatalytic: the rate at which they are generated may depend on how much of them is already present. Turing understood that, as we noted in the previous chapter, this kind of feedback can lead to instabilities that amplify small, random fluctuations and, under certain constraints, turn them into persistent oscillations. He drew the comparison with a mouse climbing up a rod that is dangling from a thread. The rod is initially in a stable equilibrium, but if 'a mouse climbs up the rod the equilibrium eventually becomes unstable and the rod starts to swing'. This is the mechanism by which symmetry is spontaneously broken.

This much sounds all rather similar to the BZ reaction; but to generate robust biological form, it is not enough for the morphogens to excite oscillations and travelling waves—they must induce *stationary* patterns in space. That is what was new and significant about Turing's paper. He showed that, in a ring of cells that produce morphogens, both travelling waves and stationary patterns in concentration might arise, depending on the conditions. He also calculated by hand* what his reaction–diffusion equations predicted for the distribution of morphogens in a flat sheet of material when they diffused under the influence of random fluctuations, and found that discrete, persistent islands of high concentration emerged with rather random, irregular shapes. Turing suspected that these 'dappled' patterns might be related to those seen on animal skins.

It is not easy to boil down Turing's complicated mathematical equations in order to extract the crucial ingredients. These were not fully clarified until 1972, when the mathematical biologist Hans Meinhardt, then at the Max Planck Institute for Virus Research in Tübingen, Germany, and his colleague Alfred Gierer identified what was involved in making a 'Turing structure'. To obtain stationary patterns, the series of chemical reactions of morphogens needs something extra beyond the autocatalysis that provokes runaway instabilities. Let us say that morphogen A undergoes autocatalysis, so that the rate of generation of A is proportional to the amount of A already present. The other vital factor is that A promotes the formation of a second compound B that *inhibits* the formation of A (Turing himself regarded this inhibitor substance as a kind of 'poison'). There is then a competition between *activation* of the autocatalytic process by compound A and *inhibition* by B. And for this to lead

---

*The difficulty of solving his equations by hand, and the approximations that this forced him to make, frustrated Turing. He suggested that a 'digital computer' might enable one to investigate a few particular cases of his theory more accurately. It was just such a device, of course, that Turing himself was trying to develop.

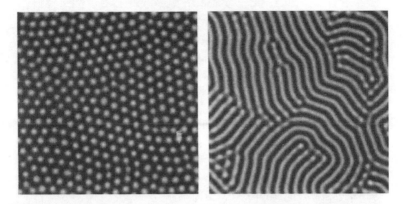

FIG. 4.2    A chemical activator–inhibitor system generates patterns of spots and stripes. The chemical composition of the medium differs in the light and dark regions, even though the molecules can diffuse freely. (Images: Jacques Boissonade, University of Bordeaux.)

to stationary patterns, the rates of diffusion of A and B must differ so that B travels more rapidly than A. This means that autocatalytic production of A may dominate in localized spots, but that it is suppressed over longer distances by B. The rapid diffusion of B ensures that it does not inhibit the generation of A over the short range, because it disperses too quickly. Then, A and B dominate in different parts of the system, and it becomes inhomogeneous. Provided that the ingredients of the reactions are constantly replenished, and the end products or wastes removed—that is, provided there is a throughput of materials, as in a continuous stirred-tank reactor, to keep the system away from equilibrium—these patches can persist indefinitely.

Meinhardt and Gierer called Turing's mechanism an *activator–inhibitor* scheme. This is one particular variety of a reaction–diffusion process. The Turing structures that the mechanism generates can now be calculated with ease on a computer. It turns out that two general kinds of pattern are formed: spots and stripes (Fig. 4.2). The pattern features all have the same width, and are spaced an even distance apart, so the patterns have a degree of regularity and a characteristic *wavelength*. We have to remember that they are not 'frozen' into the medium, like for example the polymer patterns shown in Chapter 2 (page 94): the molecules are constantly moving through the medium and reacting, and the patterns are thus *dynamic*, remaining in place only as long as the mixture is driven away from equilibrium by the input of materials and energy. They are rather like the acoustic standing waves in an organ pipe, where the variations in air density are fixed in space even though the air molecules themselves are diffusing freely.

Alan Turing would surely have continued to develop and refine his theory of stationary chemical patterns if he had not died two years after publishing his seminal paper. Having played a central part in the code-breaking work at Bletchley Park in Milton Keynes, England, during the Second World War, he was regarded as a security risk following his prosecution for homosexual activities in 1952. (Some of Turing's methods for decrypting German naval messages are still classified today.) He was ordered to take a course of 'corrective' hormone therapy, and restrictions were placed on his freedom to travel. Disgraced and depressed by this sordid treatment, in 1954 the 42-year-old Turing bit into an apple laced with cyanide.

## STRIPED PAINT

Because Turing's paper on morphogenesis was so far ahead of his time, it was more or less ignored by chemists and biologists for almost 20 years. No one knew how his rather abstract chemical scheme might be realized in practice, nor how to relate it to what (little) was known about the genetics of embryonic development. Perhaps Turing structures might turn out to be pure mathematical fictions—cute in theory, but impossible to make in the real world?

But as chemists came to grips with the Belousov–Zhabotinsky reaction in the late 1960s, the connection between this and Turing's scheme became clear. In 1971 the American chemist Arthur Winfree at the University of Chicago showed that the spiral waves of the BZ mixture could be understood to result from an activator–inhibitor process. This did not in itself make spiral waves an example of Turing structures, however—for one thing, the spirals are travelling waves, whereas Turing's patterns are stationary. The precise behaviour of such a chemical system depends on the relative rates of reaction and diffusion of the reagents. We saw that Turing patterns require an inhibitor that diffuses faster than the activator: if the reverse is true, travelling waves may be generated. And if the inhibitor, produced in response to the activator, sticks around for a long time rather than being quickly consumed by another reaction, then the system may undergo simple BZ-like oscillations. In other words, there are several different types of behaviours of which activator–inhibitor systems are capable; Turing structures are just one of these, and they appear only if the conditions are right.

One of the first theoretical schemes that linked the BZ reaction with Turing's pattern-forming process was proposed as early as 1968. In this auspicious year, Western and Soviet scientists came together for an international conference in Prague at which Belousov and Zhabotinsky's reaction was unveiled to those outside the Iron Curtain. This revelation led Ilya Prigogine and René Lefever at the University of Brussels to develop a four-step reaction mechanism, similar to the Oregonator (page 122), to explain the oscillations. But when the Brusselator, as Prigogine and Lefever's model is called, takes place in a poorly mixed solution through which the reagents diffuse at different rates, it can develop instabilities that give rise to stationary spatial patterns: Turing's spots and stripes.

This showed that Turing patterns could be generated in theory by a reaction that bore some relation to one in the real world. But it was not until more than two decades later that they were seen experimentally. They were produced in 1990 by Patrick De Kepper and his co-workers at the University of Bordeaux, using an oscillatory chemical reaction called the CIMA reaction, which De Kepper's team had devised in the early 1980s as an alternative to the BZ reaction. CIMA stands for 'chlorite, iodide, and malonic acid', the ingredients of the reaction. The oscillations can be made very striking by adding starch to the mixture, which switches between yellow and blue as it reacts with one of the intermediate compounds involved in the reaction. The CIMA reaction is closer than the BZ reaction to Turing's scheme, because it includes explicit activator and inhibitor compounds (iodide and chlorite respectively).

The key to generating Turing structures in such a mixture is to make these two compounds diffuse at very different rates. During the CIMA reaction, the compound that induces the colour change of starch (a form of iodide called tri-iodide) gets stuck to the starch molecules, which are relatively big and bulky. So when the Bordeaux group conducted the reaction in a gel,

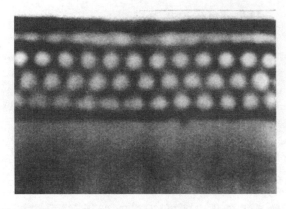

FIG. 4.3   The first observation of chemical Turing patterns. Here the dots are confined to a narrow strip where diffusing reagents meet and react. (Photo: Jacques Boissonade, University of Bordeaux.)

which is a kind of network of tangled polymer chains, the starch molecules tended to get trapped in the web with tri-iodide attached, slowing down the diffusion rate of the activator (iodide) while leaving the inhibitor unhindered. This difference meant that when the researchers introduced the various reagents in different parts of the gel and allowed them to diffuse until they met along a narrow strip, bands of stationary yellow dots appeared at this interface against a blue background (Fig. 4.3). These are genuine Turing structures.

A year later, Harry Swinney and Qi Ouyang at the University of Texas at Austin showed how to make large fields of Turing spots in a flat layer of gel infused with the CIMA mixture (Plate 5a). By changing the concentrations of the ingredients, they could transform the chemical leopard into a chemical tiger, generating Turing stripes (Plate 5b). Swinney, working in collaboration with John Pearson from the Los Alamos National Laboratory in New Mexico, also discovered a strange form of chemical spots that are rather like a hybrid of the travelling waves of the BZ scheme and the stationary patterns of Turing structures. Pearson had devised a computer model of a reaction–diffusion system that, under certain conditions, could produce spots which grow and divide in a manner superficially redolent of bacterial life. Swinney's team helped Pearson to

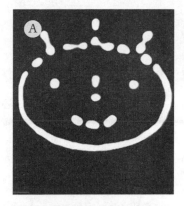 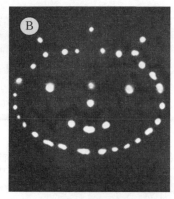

FIG. 4.4   Turing patterns can be used to create a kind of 'chemical memory.' The chemical medium, containing a light-sensitive activator–inhibitor mixture, is imprinted with a pattern by illuminating it through a mask (a). Over time, the image breaks up into isolated spots (b). (Images: Irving Epstein, Brandeis University.)

find a real chemical reaction that would generate these 'replicating spots' (Plate 6): another variant of the BZ mixture called the ferrocyanide-iodate-sulphite reaction. The spots do not appear spontaneously; they must be triggered by some perturbation, such as shining ultraviolet light onto a part of the medium. The spots grow, elongate, and split in a repetitive fashion, but if they become too populous then they annihilate one another. As in life, overcrowding leads to death.

Irving Epstein and his co-workers at Brandeis University in Massachusetts have shown that Turing structures can be used to record information; that is, to make a kind of chemical memory. They used a light-sensitive version of the BZ reaction that forms Turing structures in the dark but, because of light-induced reactions, loses them in bright light. By illuminating the mixture through a patterned mask, the researchers were able to imprint on it a pattern corresponding to the shape of the mask (Fig. 4.4a). Once imprinted, this pattern persisted indefinitely (so long as the reagents were constantly replenished) if the mixture was then uniformly illuminated with light at a finely tuned level that was sufficient to suppress pattern formation in other parts of the medium but not too bright to erase it from the parts already patterned. Over time, the pattern broke up into spots (Fig. 4.4b), recalling the way a uniform jet or cylinder of liquid fragments into a string of pearls under the influence of the Rayleigh instability.

## SKIN DEEP

Does Turing's theory answer Kipling's questions about how animals got their markings? It is certainly suggestive that the two basic patterns generated by an activator–inhibitor scheme are spots and stripes, the most common marking patterns on animals. And such a patterning mechanism would provide both economy and diversity in evolutionary terms. There is then no need for genetic encoding of a paint-by-numbers pigmentation pattern—all that is required are the basic ingredients of an inherent propensity for spottiness, which can adapt itself to bodies of any size and shape. And the same system can, with a little adjustment, be geared to stripiness instead.

So, if I were a Creator with a penchant for Gerald Manley Hopkins's 'dappled things' (though one has to be depressingly cautious about invoking deistic imagery these days), I couldn't hope for a better recipe. But as we observed with the bees' honeycomb, nature does not always make her patterns the easy way. Is this really how the zebra got its stripes?

The pelt patterns of mammals are mosaics painted in an austere palette: they are defined by hairs that are either white (unpigmented), black, brown, or yellow/orange (Plate 7). The colour of the hairs that grow from a particular patch of skin is determined by pigment-producing cells called melanocytes in the innermost layer of the epidermis. This pigment, called melanin, is a molecule that passes from the melanocyte into the hair. It comes in two forms: eumelanin is black or brown, while phaeomelanin is yellow or orange.

The pigmentation pattern, then, depends on whether pigment production in the melanocytes is turned on or off. The melanocytes are formed by differentiation of progenitor cells during the growth of the embryo: once differentiation has occurred, the pattern is 'fixed' into the skin, rather like the development of a latent photographic image in an exposed film. The question is what controls the pattern of differentiation. In other words, the riddle of animal markings is simply an aspect of the broader question that preoccupied

Turing, and Haeckel before him: as embryos grow by the proliferation and specialization of cells, what determines the organism's form?

I am going to tackle that issue head-on in the final chapter. Animal markings exemplify it in a particularly simple and beautiful way, however, because these patterns are seemingly so simple and repetitive—which immediately suggests a connection to Turing's hypothesis.

The postulate is easily stated: the pigmentation pattern defined by the melanocytes is generated by the diffusion and reaction of some chemical patterning agents—Turing's morphogens—in the embryonic epidermis. This is just what the mathematical biologist James Murray, then at the University of Oxford, proposed in the late 1970s. He argued that during the first few weeks in the genesis of an embryonic zebra or leopard or other dappled thing, activation and inhibition of chemical morphogens creates a Turing-style 'pre-pattern' that is then 'read out' by the pigment-producing melanocytes, whose location on the skin depends on the presence or absence of morphogens.

The idea sounds plausible, and it now seems likely that it is essentially correct. But to verify that directly, one would need to identify the morphogens and their disposition in a pre-pattern in the embryo. This has not yet been achieved. Murray, however, took a different approach. He asked whether, if this is indeed the way animal markings are made, it can reproduce the pattern features found in nature.

We have seen that the precise behaviour of a reaction–diffusion scheme depends on various factors, such as the relative diffusion rates of the participating molecules. Another strong influence on the selection of a particular pattern is the size and shape of the vessel in which the reaction occurs. This may seem odd: the outcome of most chemical reactions does not, after all, depend on whether they are conducted in a cylindrical test-tube or a conical flask. But Turing patterns have an intrinsic sense of scale and geometry. They are wave-like modulation of chemical concentrations with a specific wavelength, which depends (in part) on the relative diffusion rates of the reagents. There is thus an analogy with the way acoustic 'standing waves' (or *modes*) in organ pipes depend on the dimensions of the pipe.

If, for example, we try to produce a Turing structure in a container that is no bigger than the typical diameter of a single spot, we may see no pattern at all: the mixture is more or less uniform throughout, because there is not enough room to support any pattern. As the container gets bigger, patterns can appear, and the number of pattern features—the complexity of the pattern—grows. Murray has drawn an analogy with the acoustic resonances that can be excited in metal plates of different sizes: the same shape can support more complex acoustic patterns as its size increases (Fig. 4.5). The analogy is visually informative, but not exact, for sound waves are not generated by activation and inhibition, as Turing structures are.

Murray looked at whether this dependence on size and shape might plausibly account for the differences in pattern seen on animal tails. These tend to come in just two basic varieties: bands running around the tail's axis, or spots (Fig. 4.6). Either way, nearly all tails end, at their tapered extremity, in a series of bands. Murray performed calculations to see what patterns a reaction–diffusion system will generate on 'ideal' tails shaped like simple tapering cylinders, and he found that both bands and spots could be formed: bands are generated in small or narrow tails, while spots become possible in wider tails patterned by an identical process (Fig. 4.7). A transition from bands to spots may take place between the narrow tip and the broader root of the tail, as sometimes seen for cheetahs and leopards.

It is important to remember that these patterns are laid down in the embryo, and so it is this, rather than the size and shape of the adult animal, that matters. For example, while the adult genet and leopard have tails of similar shape, the embryonic genet has a thin tail of nearly uniform diameter, while the embryonic leopard has a shorter, tapered tail. This means that the genet's tail becomes banded, while the leopard's is mostly spotted.

Murray also showed that a simple stripe-forming Turing process* could account for the chevron pattern, known as scapular stripes, found at the junction of leg and body in zebras (Fig. 4.8). In the roughly cylindrical legs, stripes become circumferential bands, as they do in animal tails.

If the reaction–diffusion system that creates these pigmentation patterns is essentially the same for all mammal species, generating much the

---

*It is actually not terribly easy to generate stripes in Turing-type reaction–diffusion models; they have a tendency to break up into spots. Murray assumed that stripes could survive in his model, but typically this requires some special constraints—for example, if the rate of autocatalytic production of the activator has an upper limit, so that it can't go on 'feeding back' indefinitely.

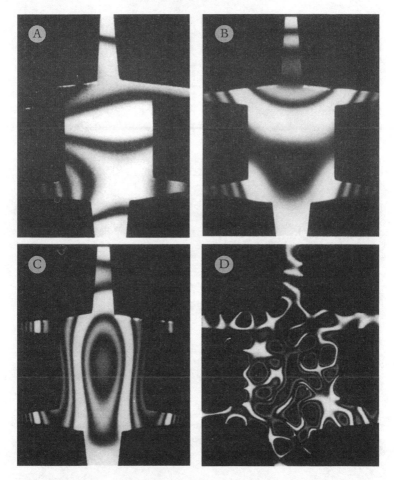

FIG. 4.5 The patterns formed by activator–inhibitor systems depend on the system size: larger systems can support more 'modes', and so the patterns are more complex. This is analogous to the complexity of standing-wave vibrations excited by sound waves in plates of different sizes. Here are the acoustic oscillations in sheets shaped to mimic the body surfaces of animals. The excitation frequency increases from a to d, which is equivalent to increasing the size of the plate, since it means that more wavelengths can 'fit' into the space available. (Photos: James Murray, University of Washington, Seattle.)

FIG. 4.6 Animal tails may be marked with either spots or bands, but bands always appear as the tail tapers towards the end, as seen here for a Geoffroy's cat (a) and an ocelot (b).

FIG. 4.7 The patterns produced on tapering cylindrical 'model tails' by an activator–inhibitor scheme depend on the tails' size and shape. Small cylinders support only bands (stripes) (a), whereas spots appear on larger tails (b). On a slowly tapering tail, the transition from spots to bands is more clear (c). (After Murray, 1990.)

same size of pattern features in the embryo (an assumption that seems reasonable but is not proven), then we might also expect issues of timing during embryogenesis to exert a dominant influence on the resulting pattern. For example, small animals with short gestation periods should have less complex pelt patterns than larger animals, because their smaller embryos may hold fewer pigmented patches. On the whole this seems to be the case: most dramatically, the honey badger and the Valais goat are boldly divided into a white half and a black half.

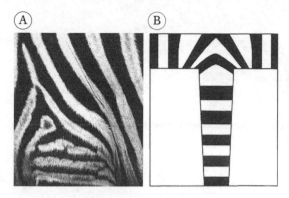

FIG. 4.8 The scapular stripes of a zebra, where the leg meets the torso, form a kind of chevron pattern (*a*), which is reproduced in an activator–inhibitor system confined within a similar geometry (*b*). (Photo: a, Thiru Murugan.)

A more finely tuned question of timing may account for pattern variation between zebra species. The common zebra *Equus burchelli* has about 26 broad stripes on its body from nose to tail, whereas the imperial zebra *Equus grevyi* has closer to 80, and the mountain zebra *Equus zebra* around 55 (Fig. 4.9). In 1977 the zoologist Jonathan Bard at Edinburgh University proposed that these differences arise because differentiation of melanocytes happens at different times for each species, three to five weeks into the gestation period. As the patterning is fixed after just 21 days for *E. burchelli*, it produces fewer stripes because the embryo is smaller. The same process, giving a similar stripe spacing, creates a larger number of stripes in the larger *E. grevyi* embryo after five weeks (Fig. 4.9*c–e*). Note that these stripes appear in the embryos around the backbone first, because the cells that produce melanin are formed here and gradually migrate towards the belly and chest as the embryo develops. They do not always go the distance: the imperial zebra has a white belly, and so of course do many other pigmented animals, such as cats and dogs.

Surprisingly, these animal-skin patterns should be less complex for very large animals as well as for small ones. This is because, as more and more features can be supported on the embryo's surface, they start to merge and the dividing lines get squeezed out. Murray suggested that this is why giraffes have very closely spaced dark spots, with narrow boundaries, and why elephants and hippopotami have no markings at all: more, in this case, is less.

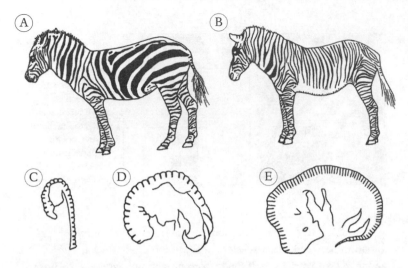

FIG. 4.9   (Below) The common zebra *Equus burchelli* (a) and the imperial zebra *Equus grevyi* (b) have different numbers of stripes. This is thought to be because the striped 'pre-pattern' laid down on the embryo happens at different times: after 21 days for *Equus burchelli* (c), but after five weeks for *Equus grevyi* (e). By the time the *Equus burchelli* embryo has grown to a comparable size to *Equus grevyi* when the latter acquires its pattern, its stripes are therefore wider (d). (After Murray, 1989.)

Murray's activator–inhibitor model for giraffe patterns produces blob-like spots with rounded edges (Fig. 4.10b)—merely bloated, crowded versions of the leopard's spots. Real giraffe markings don't look much like this. They are instead irregular polygons with more or less straight sides, so that the light boundaries form a reticulated web rather like the cracks in dried mud (Fig. 4.10a). It is not obvious how a normal Turing-style activator–inhibitor scheme can achieve something like this. But Hans Meinhardt and his colleague André Koch have shown that the pattern can be created by a variant in which the activator and inhibitor diffuse at similar rates. This, as I mentioned above, produces not stationary blobs but travelling waves. These can become translated into a fixed pattern by assuming that, if the activator concentration exceeds a certain threshold at any point in space, this throws a kind of biochemical switch that triggers or suppresses pigment production. In the model of Meinhardt and Koch,

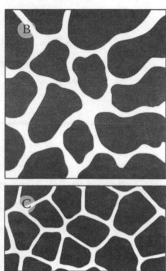

FIG. 4.10  The pelt pattern of a giraffe consists of very large features that almost merge, with narrow boundaries (*a*). A simple activator–inhibitor model produces large blotches that are only crude approximations to this pattern. A more sophisticated model in which travelling waves of activator and inhibitor throw a biochemical 'switch' that triggers pigment production generates more realistic polygonal shapes (*c*). (Photo and images: *a*, Michael and Simon Ball; *b*, After Murray, 1990; *c*, After Koch and Meinhardt, 1994.)

once the switch is flipped, it stays that way, even if the level of activator subsequently falls below the threshold again.

Within this model, the production of the activator compound is initiated at random points, triggering chemical waves that spread until they meet another, at which point they annihilate one another. These boundaries between each wave-generating domain are straight lines, and so the system gets separated into polygonal domains within which pigment production is switched on, separated by a net of unpigmented boundaries (Fig. 4.10*c*). This bears a closer resemblance to the pattern on real giraffe pelts.

The same basic model also enabled Meinhardt and Koch to obtain a better approximation to the leopard's spots, which are not mere blobs of pigmented hairs but are shaped like rings or crescents. But that might not have satisfied Kipling, who astutely noted that the leopard's spots are in fact neither blobs

FIG. 4.11   The 'spots' of a leopard (a) and a jaguar (b) are in fact rosette patterns, with slightly darkened interiors enclosed by broken rings of black. Those of the jaguar have a polygonal shape, and contain some black spots in the centres too. These patterns evolve as the young cats grow: a jaguar, for example, has simple spots at five weeks, which become irregular rings at three months (c). (Photos courtesy of Sy-Sang Liaw, National Chung-Hsing University, Taiwan, and Philip Maini, University of Oxford; from Liu, Liaw and Maini, 2006.)

nor crescents, but rosettes—rings of dots, resembling the imprints of finger-tips (Fig. 4.11a). The dots are black, but they enclose a space that is tan-coloured, while the intervening pelt between rosettes is lighter. Superficially the jaguar is spotted like the leopard; but a more careful inspection shows that the markings are different again, being polygons not unlike those of the giraffe, except that the pigmented patches are not uniform, but have a black outline and a tan inner core decorated with dark spots (Fig. 4.11b). Moreover,

FIG. 4.11 (Continued).

the patterns in both these wildcats alter as the animals grow (Fig. 4.11c): they both begin as simple dark spots (called flecks), before evolving into the more elaborate pattern features in the adult beasts. A Turing-type model that produces spots or even crescents may look appealing, but should we be content with this snapshot of a rather approximate similarity?

FIG. 4.11 (Continued).

We needn't be. Philip Maini of Oxford University and his co-workers have shown that more realistic versions of the leopard's spots can be produced in an activator–inhibitor system if one makes two assumptions: that reactions between the activator and inhibitor morphogens have a particular mathematical form, and that the rates at which these reactions occur may change over time. In this case the system may generate spots that evolve into rings, which then themselves break up into irregular rings of spots, or rosettes (Fig. 4.12*a*). By tweaking the parameters of the model a little, Maini and colleagues could obtain jaguar-like spots in polygonal rings, rather than leopard-like rosettes, in the final stage (Fig. 4.12*b*). In both cases, the evolution of the patterns follows much the same course as that seen in practice in the growing cats.

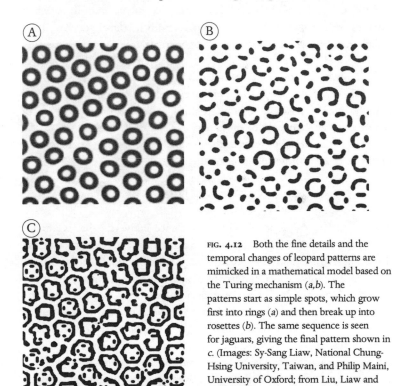

FIG. 4.12 Both the fine details and the temporal changes of leopard patterns are mimicked in a mathematical model based on the Turing mechanism (*a,b*). The patterns start as simple spots, which grow first into rings (*a*) and then break up into rosettes (*b*). The same sequence is seen for jaguars, giving the final pattern shown in *c*. (Images: Sy-Sang Liaw, National Chung-Hsing University, Taiwan, and Philip Maini, University of Oxford; from Liu, Liaw and Maini, 2006.)

None of this proves that the theoretical reaction–diffusion scheme matches the biochemical process occurring in the pelts of these animals. But it shows that the Turing model is able in principle to reproduce realistic marking patterns with a rather limited set of ingredients. In other words, the job might or might not be done this way, but it certainly *could* be.

Maini's colleague Sy-Sang Liaw at the National Chung-Hsing University in Taiwan believes that Turing patterning might operate in a quite different kind of creature, the lady beetle, or ladybird as they are known in the United Kingdom, whose red and black wings have earned them an affection almost unique among insects. These 'wings' are in fact hard coverings called elytra which protect the more fragile hindwings and the soft abdomen beneath. These patterns appear over a matter of minutes or hours after the adult emerges from its pupal stage, and are usually made up of black spots—although stripes are found on some lady beetle species. Liaw and his colleagues have calculated how the features of a Turing structure will be distributed over a curved shell that approximates the shape of the lady beetle's hard wings, and they find that the precise positions of the spots may depend on the degree of curvature, the shape of the boundary, and the diffusion rates of the morphogens assumed to exist in their model. The resulting patterns can mimic several of those found in nature (Fig. 4.13).

## HARD STUFF

There is at face value no obvious discordance between what we might regard as D'Arcy Thompson's caricature Darwinist who sees a pattern like the zebra's stripes and cries 'Adaption!', and the modern morphologist who shows how such a pattern might arise from chemical processes. The one 'explains' the persistence of a pattern in the population; the other shows how, on an individual creature, the features are put in place. The question remains, however, of whether nature really does paint in this spontaneous manner. And if so, how much latitude does she have to ring the changes? I shall return to these points shortly.

First we must give that enthusiastic Darwinist pause for thought. Exhibit A: a handful of mollusc shells. As canvases for natural patterns, these

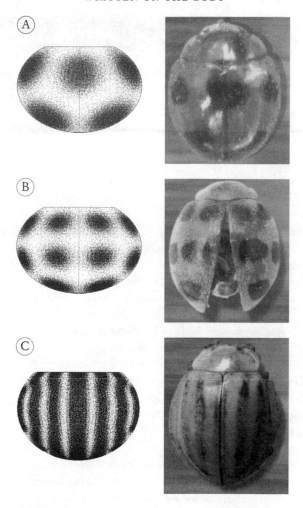

FIG. 4.13 Modelled and real patterns on various species of ladybird: *Platynaspidius quinquepunctatus* (a), *Epilachna crassimala* (b), and *Macroilleis hauseri* (c). (Photos and images: Sy-Sang Liaw, National Chung-Hsing University, Taiwan; from Liaw *et al.*, 2001.)

FIG. 4.14 Shell patterns on molluscs can show large variations even among members of the same species. The shells of garden snails, shown here, carry stripes of quite different widths. (Photo: Hans Meinhardt, Max Planck Institute for Developmental Biology, Tübingen.)

mineralized tissues are hard to surpass. The arrangements of pigments on their surfaces are often exquisite and bewilderingly diverse. But attempting to account for this profuse invention in terms of its adaptive benefits— whether as camouflage, danger signals, species recognition, or whatever we choose from the usual explanations for surface patterning in biology— will not get us very far. Many molluscs live buried in mud, where their elaborate exterior designs are totally obscured. Others cover their shell markings with an opaque coat of organic tissue, as if embarrassed by their own virtuosity. And individual members of a single species may exhibit such personalized interpretations of a common theme that you have to doubt whether they would recognize one another as members of the same population (Fig. 4.14). It appears that these patterns can serve no evolutionary purpose whatsoever.

The only conclusion one can draw is that nature has an irrepressible impulse to create patterns, and does so even when there is no 'need' for them. It is tempting to come over all misty-eyed, if not downright mystical, about this, which of course is just what we ask for whenever we anthropomorphize nature. Nevertheless, it is striking to discover that nature is not, as some biologists might once have had us believe, the austere censor, ruthlessly stripping away all unnecessary detail and ornamentation like a Modernist zealot. Some patterns just 'happen', and it is fruitless to seek a reason in terms of *function*. That still surprises some biologists, even though they know very well that Darwinism does not insist that every feature of an organism must be adaptive. But I hope this series of books will convince you

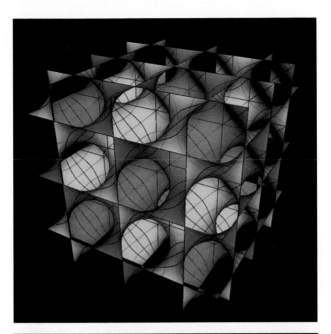

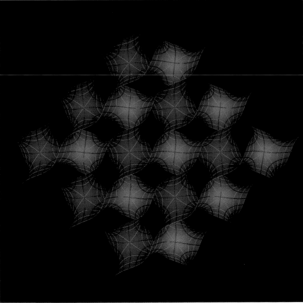

**PLATE 1:** The D (diamond) and G (gyroid) minimal surfaces.

(Images: Matthais Weber, Indiana University)

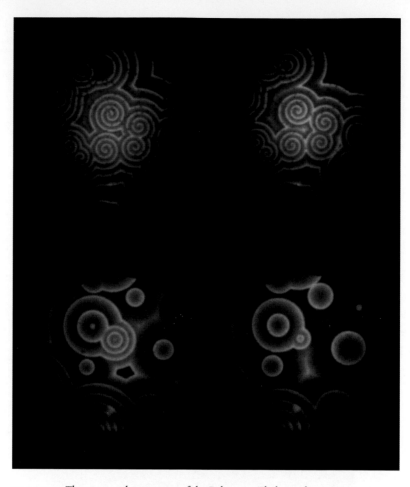

**PLATE 2:** The spectacular patterns of the Belousov–Zhabotinsky reaction.

(Photos: Arthur Winfree, University of Arizona)

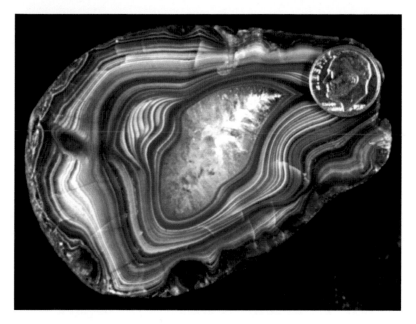

**PLATE 3:** The concentric patterns of banded agate are created by a periodic cycle of precipitation as the mineral forms in the Earth.

(Photo: Peter Heaney, Princeton University)

PLATE 4: Chevron patterns in a colony of the bacterium Escherichia coli grown in a plate of gel. The patterning is a collective response to the adverse conditions, such as a scarcity of nutrients, which induce the bacteria to aggregate by chemical signalling and movement (chemotaxis).　　(Photo: Elena Budrene, Havard University)

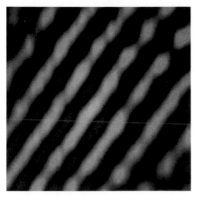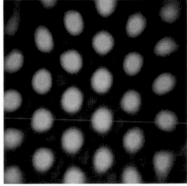

**PLATE 5:** Turing patterns in a chemical medium. These patterns arise spontaneously from competition between a localized autocatalytic (self-amplifying) chemical reaction and the long-ranged diffusion of a substance that inhibits the reaction. The colours (which are real) correspond to regions of different chemical composition. The patterns are static, but a switch from stripes to spots can be induced by changing the ratios of ingredients in the mixture.

(Photos: Harry Swinney, University of Texas at Austin)

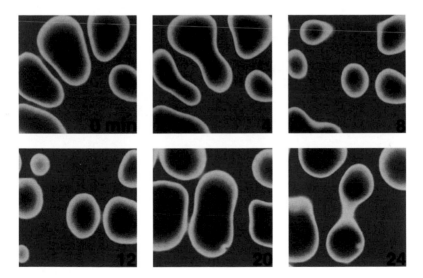

**PLATE 6:** Replicating spots in an activator-inhibitor chemical reaction. These spots grow and divide. When too many are grouped in one region, they may 'die' through overcrowding.                    (Photos: Harry Swinney, University of Texas at Austin)

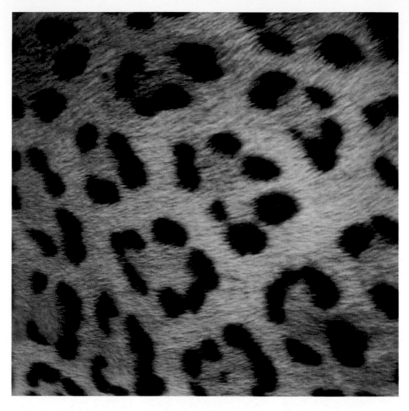

PLATE 7: The characteristic markings on the pelts of great cats (here a Paraguayan jaguar) are produced by epidermal cells that generate hair-colouring pigments.

PLATE 8: The stripes on an angelfish are continually evolving as the creature grows.

(Photos: (A) Jenny Huang; (B) cadmanof50s)

PLATE 8: Some of nature's most spectacular patterns are found in the kaleidoscopic designs of butterfly wings.

(Photos: H. Frederik Nijhout, Duke University, North Carolina)

that it is not surprising at all—for many of the patterns of biology can be found mirrored in non-living systems, which cannot, unless you are a certain kind of animist or theist, display any purpose at all. The stripes and other, sometimes rather baroque designs on mollusc shells may superficially resemble those seen on animal skins, and they may well share some of the same formative principles; but they are not subject to selective pressure and so are apparently not there for any Darwinian reason. Here, according to Hans Meinhardt, nature is 'allowed to play'.

Shell pigmentation patterns are not like most animal markings in at least one important respect, however: they are not laid down all at once. For a shell gets bigger by accreting calcified material continuously along its outer rim, and so the shell's form and appearance adds up to a *historical record* of this essentially one-dimensional patterning process, rather as a contrail is the historical trace of a jet's progress across the sky. The pigmentation pattern on the shell's surface is a trace of the way the distribution of pigment has varied over time along the rim of the shell. Thus, stripes that run parallel to the growth axis are superficially similar to those that run perpendicular to it (Fig. 4.15*a*), but they speak of quite different patterning processes: the first in which bursts of pigmentation occur all along the rim, followed by spells of unpigmented growth, the second in which a spatially periodic pattern of pigmented sections on the growing edge remains constantly in place as the shell grows. Stripes that run at an oblique angle to the growth direction, meanwhile, are manifestations of travelling waves of pigmentation that move along the rim as growth proceeds (Fig. 4.15*b*). These different types of markings can therefore be rationalized as uniform on-off oscillations in pigmentation, or BZ-style travelling waves, or Turing-style stationary patterns.

Meinhardt has shown how reaction–diffusion schemes can account for just about any kind of pattern that molluscs have been found to concoct. Bands around the shell's axis, for instance, are produced by a stationary pattern of spots, the one-dimensional equivalent of Turing's leopard. The widths and spacing of the spots can be acutely sensitive to the exact values of the model parameters, such as the relative diffusion rates of activator and inhibitor (Fig. 4.16). So, differences between members of the same species, as seen in Fig. 4.14, might be the result of slight variations in the growth conditions, such as temperature, which alter the diffusion rates.

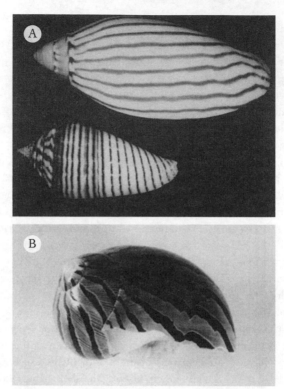

FIG. 4.15 Stripes that run parallel to, perpendicular to, and oblique to the axis of a shell reflect quite different patterning mechanisms. In the first case (*a, top*) the stripes reflect a patterning process that is uniform in space but periodic in time. In the second (*a, bottom*) the pattern is periodic in space but uniform in time. In the third case (*b*), it results from travelling waves, periodic in both time and space. (Photos: Hans Meinhardt, Max Planck Institute for Developmental Biology, Tübingen.)

Because the shell patterns are traces of the growth history, the same basic patterning mechanism can give rise to patterns that look rather different simply because of differences in shell shape. Thus, for example, the equally spaced spots of pigmentation that create bands in a spiralling shell like that shown in Fig. 4.15a would generate spoked patterns on conical shells (Fig. 4.17). In the latter case, the overall length of the perimeter is steadily increasing. This means that the spots move

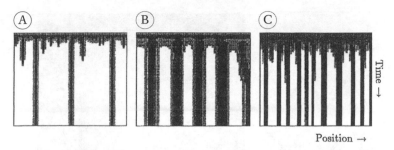

FIG. 4.16 Stripes perpendicular to the growth edge of the shell are the result of one-dimensional patterning at the rim. The pattern gets 'pulled' into stripes as the shell's rim advances (*a*). If the activator in this reaction scheme diffuses more rapidly, the stripes are broader (*b*). If the concentration of activator rises until it 'saturates', the spacing of the stripes becomes irregular (*c*). (Images: Hans Meinhardt, Max Planck Institute for Developmental Biology, Tübingen.)

FIG. 4.17 When a shell's growth edge traces out a cone, a one-dimensional periodic pattern at the edge becomes a radial spoke pattern. As the edge gets longer, new pattern features may appear in the spaces between existing spokes (see right). (Photo: Hans Meinhardt.)

progressively further apart, and it becomes possible, at certain stages during growth, to incorporate a new spot in between them (recall that the average spot spacing stays constant, depending on the relative diffusion rates), which then grows subsequently into a new spoke.

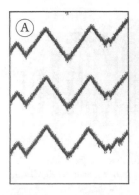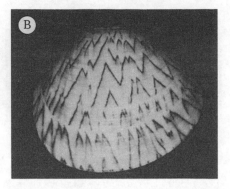

FIG. 4.18  Annihilation between one-dimensional travelling waves leads to V-shaped patterns (*a*), as seen on the shell of *Lioconcha lorenziana* (*b*). (Image and photo: Hans Meinhardt.)

When Meinhardt's model produces travelling waves, the result is a series of oblique stripes on the shell as the wavefronts move along the rim. Sometimes these fronts move in just one direction, as in Fig. 4.15*b*. But just as in two dimensions a travelling wave tends to move away from its source as a ring-shaped wavefront, creating target patterns, so in the one-dimensional system of the shell rim a wave may be dispatched from its source in both directions. The time-history recorded on the shell is then an inverted V, with the apex pointing away from the rim (Fig. 4.18). When two of these wavefronts meet, they annihilate one another like the waves of the BZ reaction, with the result that two diagonal stripes converge in a V with the apex pointing *towards* the rim. Both features can be seen on real shells.

Occasionally one finds a shell that seems to have had a change of heart during growth: it displays a beautiful pattern that abruptly turns into something else (Fig. 4.19). In such cases, activator–inhibitor schemes may account for both patterns—but why the switch? It seems likely that the shell has experienced a shock because of some external agency—an environmental disturbance, say, when the location became suddenly warmer or drier or food more scarce—and that this knocked the biochemistry of the mollusc off balance. (Remember that it is the soft and squishy creature within the shell that is responsible for supplying the materials and energy needed to

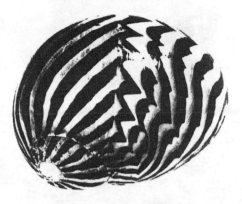

FIG. 4.19 Sudden changes in environmental conditions can reset the patterning process on shells, creating abrupt discontinuities in the pattern. (Photo: Hans Meinhardt.)

construct the hard coating.) Such a disturbance can 'reset the chemical clock', and the new pattern, arising in a new environment, might actually be quite different from the old one. Like all artists, molluscs need to be left alone in comfort to do the job well.

## FISHY EVIDENCE

Biologists can be hard to convince. One might think that the striking similarities between the patterns on animals and shells, and those that can be generated by Turing's activator–inhibitor scheme and its variants, would be enough to prove that this is how zebras get their stripes and so on. But as long as such arguments rely only on visual comparisons, the evidence can never be more than circumstantial. We can, however, do rather better than that.

In 1995 the Japanese biologists Shigeru Kondo and Rihito Asai at Kyoto University looked at the stripe markings of angelfish, which exhibit some of the most beautiful and eye-catching patterns in the ocean (Plate 8). It is common knowledge that stripes can be manufactured with an activator–inhibitor scheme, but what distinguishes the angelfish is that the stripes of some of its species, such as those of *Pomocanthus imperator* (emperor) or *P.*

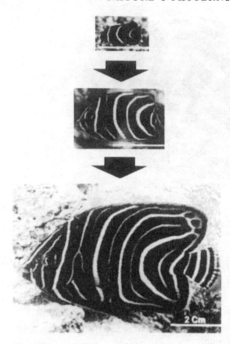

FIG. 4.20 As the angelfish grows, its stripes maintain the same width and spacing—so the body acquires more of them. This contrasts with the patterns on mammals such as the zebra or cheetah, where the basic patterns are laid down once for all at the embryonic stage and then expand like markings on a balloon. (Photo: Shigeru Kondo, Kyoto University.)

*semicirculatus*, do not seem to be permanently imprinted into the fishes' scaly skin at an early stage of development, as they are on zebras and cheetahs. Rather, the pattern continues to evolve as the fish grow. Or rather, we should say that *the pattern stays the same* but the fish changes, so that small fish have fewer stripes, but acquire more as they grow (Fig. 4.20). These changes happen quite abruptly. When they are less than 2 cm long, young *P. semicirculatus* have about three stripes. These get a little wider and further apart as the fish get bigger, but once they reach about 4 cm in length, a new stripe emerges in between the existing ones, and they revert back to the spacing seen in the younger fish. This process repeats when the body reaches 8–9 cm. In contrast, the features on, say, a giraffe just get bigger as the body expands, like a design on an inflating balloon.

This must mean that the angelfish stripes are being actively sustained during the growth process—the reaction–diffusion process is *still going on*, and the pattern it produces adjusts to changes in the dimensions of the 'container'. One might suspect that, if the fish were able (which they are not) to grow to the size of a football, the effect of scale might even lead to a completely different sort of pattern. In any event, Kondo and Asai were able to reproduce the behaviour of the stripes in a theoretical model of an activator–inhibitor scheme. What made this demonstration all the more compelling was that the researchers were able to duplicate the details of the process by which new stripes appear.

Adult emperor angelfish, for instance, have stripes that run parallel to the head-to-tail axis of the body. (The young fish, in contrast, have concentric semicircular stripes like those of *P. circulatus*—this reorganization of the pattern is striking in itself.) The spacing of these parallel stripes remains uniform, and so the number of stripes is proportional to adult body size. New stripes grow from a forking or bifurcation of existing stripes, which runs along the body in a kind of unzipping process to split a single stripe into two (Fig. 4.21). The reaction–diffusion model posited by Kondo and Asai mimicked this behaviour exactly. It also captured the more complex behaviour of stripe branching points near the top and bottom of the body (dorsal and ventral regions) (Fig. 4.22). The Japanese researchers think that in living systems of this sort, where the patterning process is still active rather than having been frozen in place during

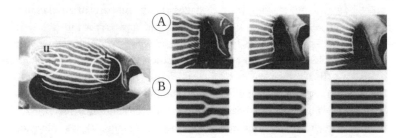

FIG. 4.21 The 'unzipping' of new stripes in *Pomacanthus imperator* (*a*: region I circled on the left) can be mimicked in a Turing-type process (*b*). (Photos and images: Shigeru Kondo.)

189

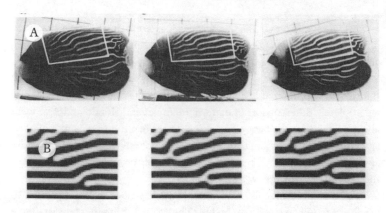

FIG. 4.22 Complex pattern reorganization in the dorsal and ventral regions of *Pomacanthus imperator* (a) is also captured by the Turing model (b). (Photos and images: Shigeru Kondo.)

embryogenesis, it might be easier to identify the activator and inhibitor molecules that are at work.

Kondo and his co-workers have discovered an even more striking example of what appears to be an active reaction–diffusion process taking place on an animal's skin. In 2003 they saw travelling waves of pigmentation, like the chemical waves of the BZ reaction, moving slowly across the skin of hairless mice. Like many other creatures, mice grow their hairs cyclically: at regular intervals a new hair grows to push the old one out of the follicle, leading to moulting. This cycle is not in step everywhere on the body, however, but reaches different stages in different parts of the skin in a way that suggests it might be controlled by some kind of reaction–diffusion wave. Kondo and colleagues found that these waves of hair development became visible in mice with a genetic mutation that leads to hairs being discharged from the follicle early, just after pigment begins to accumulate. Because of this premature hair loss, the mutant mice look essentially hairless—but the pigmentation that builds up in the follicle nevertheless colours the skin. They found that one mouse appeared spontaneously in a colony of the hairless mutants that had travelling stripes of pigmentation on its skin. The researchers identified the genetic mutation responsible for this behaviour and transferred it into other mice, which produced various types of

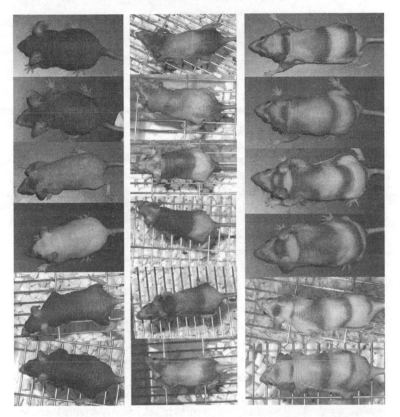

FIG. 4.23 Travelling waves of pigmentation on naked mice, evolving during the first eight months after birth. For the first 30 days or so, the entire body switches between being pigmented and unpigmented (left column), but by 210–240 days (right column), the pattern has evolved into symmetrical bands. (Photo: Shigeru Kondo, from Suzuki et al., 2003. Copyright 2003 National Academy of Sciences.)

pigmentation cycles. Typically, the mice's skin colour oscillated uniformly all over the body for about a month after birth, but within seven to eight months this changed into moving stripes of pigment (Fig. 4.23). The patterns were symmetrical along the head-to-tail axis, and Kondo and colleagues remarked that they looked rather like what one would expect from a pair of travelling-wave sources located in the armpits of the mice.

FIG. 4.24 The banded pigmentation of porcupine quills. (Photo: Tim Parkinson)

The Japanese researchers suspect that waves of pigmentation on animal skins might be common. If the colouring waves oscillate with a shorter period than the cycles of hair formation, then the same hairs experience periods of pigmented and unpigmented growth before they grow out. As a result, they should have dark and white bands. That is just what is found, for example, in some mice and domestic cats, and most strikingly, in the quills of porcupines (Fig. 4.24).

The case for patterning by Turing-like mechanisms here looks fairly compelling, but the clinching piece of evidence would be the identification of a genuine morphogen—a chemical compound that diffuses through the tissues, switching on pigmenting genes as it goes. So far, nothing of that kind has been identified for animal marking patterns. But we will see in the final chapter that there is now good evidence for diffusing morphogens of a more general sort that control development and growth in embryos.

## ON THE WING

Since Liesegang bands were the closest D'Arcy Thompson came to finding a reaction–diffusion system, we can understand that he seemed determined to get as much mileage as he could out of this barely understood patterning process. He proposed, albeit tentatively, that something analogous was at play not only in the formation of stripes and whorls on cats and birds but also in the wing patterns of moths and butterflies. This was not his idea, in fact: the German zoologist W. Gebhardt suggested it in 1912. In particular, the concentric circles of eyespot patterns on the wings of species such as the emperor moth offered themselves as highly suggestive analogues of Liesegang's rings.

But compared with the spots and stripes of mammal markings, the patterns on butterflies and moths are daunting in their richness (Plate 9). For one thing, they are multicoloured: there is more on this palette than the yellows and browns of melanins (although butterflies do use those pigments too). Some of the blues are derived from plant pigments; other colours are made not by pigments molecules at all but by the scattering of light from arrays of tiny ridges on the wing scales (see page 93). Most greens and blues are made this way, and it results in a silky, iridescent appearance. These colours are arrayed in pointillist fashion, each of them assigned to tiny scales that overlap on the wing surface like shingles. And if the palette is rich, so too are the designs, each of them built up from permutations of basic elements that are reproduced in each wing with perfect symmetry.

Is there any underlying system to this profusion, or is the wing a blank sheet on which nature can doodle as she pleases? That is what B. N. Schwanwitsch and F. Süffert asked in the 1920s, when they reduced all the wing patterns of butterflies and moths into a unified scheme known as the nymphalid ground plan. This was supposed to represent the basic template for patterning, a kind of Platonic design from which a huge variety of real patterns could be derived by selecting, omitting, or adapting the fundamental elements. The two zoologists developed their systems independently, but they showed impressive agreement (Fig. 4.25). The basic pattern elements are series of spots, arcs, and bands that cross the wings from the top (anterior) to the bottom (posterior) edges. These top-to-bottom features are called symmetry systems because they can be

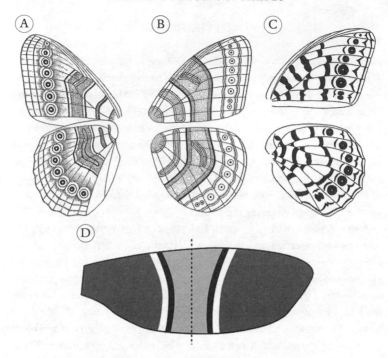

FIG. 4.25 The nymphalid ground plans of Schwanwitsch (a) and Süffert (b) represent the 'Platonic' ideal of all butterfly and moth wing patterns. They both contain features from which almost all observed patterns can be derived. An updated version of the ground plan by Frederik Nijhout (c) takes more explicit account of the wing veins. The bands running from the top to the bottom of the wings are called the central symmetry system (d). The mirror-symmetry is denoted by a dashed line. (Images: a-c, H. Frederik Nijhout, Duke University, North Carolina.)

regarded as bands or sequences of discrete elements that form approximate mirror images either side of their centre (Fig. 4.25d). Even the most complicated of wing patterns can be generally broken down into some combination of three or four symmetry systems arranged side by side, although the elaborations and modifications can sometimes make the ground plan hard to discern beneath it all.

The existence of an idealized ground plan captured the interest of the Russian-American writer Vladimir Nabokov, who, before achieving fame and notoriety with novels such as *Lolita* (1955), acted as curator at Harvard University's Museum of Comparative Zoology. In the 1940s Nabokov became something of an expert on butterflies, publishing articles in leading scientific journals and having several new species named after him. But, like D'Arcy Thompson, he was not wholly satisfied with Darwinian explanations for biological patterning: he did not believe, for example, that it was enough to explain the resemblances between two quite different species of butterfly on the grounds of mimicry that enhanced their prospects for survival.* Nabokov suspected that there is another kind of imperative at work, one that restricts the patterns on which nature could draw. In this picture, an appeal to *function* is unnecessary in accounting for the resemblances, and perhaps even misleading. Rather, Nabokov felt that there are more fundamental rules governing the pattern structure, which are hinted at in the nymphalid ground plan.

After all, this mimicry sometimes went much further than it had any adaptive reason to, Nabokov argued:

> When a certain moth resembles a certain wasp in shape and color, it also walks and moves its antennae in a waspish, unmothlike manner. When a butterfly has to look like a leaf, not only are all the details of a leaf beautifully rendered but markings mimicking grub-bored holes are generously thrown in. 'Natural selection,' in the Darwinian sense, could not explain the miraculous coincidence of imitative aspect and imitative behavior, nor could one appeal to the theory of the 'the struggle for life' when a protective device was carried to a point of mimetic subtlety, exuberance, and luxury far in excess of a predator's power of appreciation. I discovered in nature the non-utilitarian delights that I sought in art. Both were a form of magic, both were a game of intricate enchantment and deception.

This all sounds a little too mystical, or at least whimsical, for some tastes: Nabokov has been accused of advocating a departure from Darwinism that verges on creationism. And it is true that his reasoning was not always particularly scientific—he was influenced by the belief in a creative force

---

*The imitation of the wing markings of a poisonous butterfly species by a harmless one, so as to fool predators, is known as Batesian mimicry after the British zoologist Henry W. Bates.

of nature espoused by the French philosopher Henri Bergson. But we can now see that he was on to something, for as we have already witnessed in the patterns on mollusc shells, nature does indeed have a creative aspect in the way she may sometimes dabble with an intrinsic pattern-forming process regardless of any biological 'meaning'. To Nabokov, butterfly wing patterns are not the result of random doodles that chance upon an evolutionarily successful design; they are 'variations on a theme', determined by the constraints of a pattern-forming process that is delineated by the nymphalid ground plan. As Victoria Alexander of the Dactyl Foundation for the Arts and Humanities in New York puts it,

> Although Nabokov's obsession with the dynamical nature of patterns may have seemed eccentric to other lepidopterists in the 1940s, it is now clear that Nabokov was beginning to sketch out a theory of spontaneous pattern formation that was not fully articulated until the 1950s when Alan Turing published 'The Chemical Basis for Morphogenesis'.

Indeed, there is now good reason to think that butterflies and moths do make use of the system that Turing sketched out in which the wing pattern is laid down during pupation by diffusing morphogens that programme certain cells to develop into scales of a particular colour.

How exactly does this happen? Schwanwitsch noticed one key point: most wing patterns are strongly influenced by the veins that provide the framework for the wings, rather like the lead cames that hold the panels of stained-glass windows. In some species, in fact, the wing pattern simply outlines the veins with coloured borders. In general the stripes that cross the wing from top to bottom are offset where they cross a vein. Schwanwitsch called these offsets dislocations, by analogy with the dislocations of sedimentary strata where they are cut by a geological fault. H. Frederik Nijhout of Duke University has proposed an updated version of the nymphalid ground plan which features these dislocations at veins much more prominently (Fig. 4.25c). In fact, it seems that the wing-spanning bands are not true bands at all but merely independent stripes within each veined panel that happen to more or less line up. Nabokov noted this, calling the bands 'pseudo lines', or even 'manmade', in the sense that our perception constructs them out of segments.

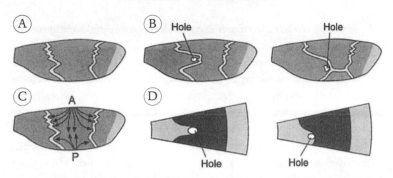

FIG. 4.26   The moth *Ephestia kühniella* has a central symmetry system defined by two light bands (*a*). Kühn and von Engelhardt investigated the formation mechanism of these bands by cauterizing holes in pupal wings and observing the effect on the pattern (*b*). They hypothesized that disruptions can be explained by invoking 'determination streams' of some chemical agent (morphogen) issuing from centres located on the anterior (A) and posterior (P) edges of the wing (*c*). There is some correspondence between the pattern boundaries in these experiments and those generated in an idealized model in which a reaction–diffusion system switches on genes that fix the pattern (*d*).

The mirror-image symmetry systems of the ground plan suggest that the pigmentation patterns might be generated by diffusion of some patterning compound—a morphogen—from their centre. This is particularly evident in Nijhout's ground plan, which shows segments of roughly circular 'wave-fronts' within each wing panel. The idea of pigmentation by diffusion was suggested in experiments performed by the zoologists Alfred Kühn and A. von Engelhardt in 1933. They suspected that the bands running from top to bottom of the wings of the moth *Ephestia kühniella* (dislocated by the wing veins) were produced by some kind of pigmentation signals that passed from cell to cell. So they looked at the effect of disrupting this signal by cauterizing small holes in the wings of the moths during the first day after pupation. As they anticipated, the bands became deformed around the holes (Fig. 4.26), as though blocked by these obstacles. They proposed that the bands represent the front of a propagating patterning signal—a 'determination wave'—issuing from points on the anterior and posterior wing edges.

Clearly this anticipates the idea of a diffusing chemical morphogen. Jim Murray and Frederik Nijhout have both devised reaction–diffusion models that can account for the shape of the deformed bands (Fig. 4.26*d*) on the assumption that the pigmentation pattern is triggered when the concentration of the morphogen exceeds a certain threshold and throws a biochemical switch that induces colouration.

The idea that patterning is orchestrated by morphogens underpins all work on butterfly wing patterns today. It is not enough, however, to explain them by postulating certain arrangements of morphogen sources—places where these chemical agents are generated. As Gierer and Meinhardt pointed out in 1972, to form stable patterns we also need sinks, where morphogens are consumed. These sources and sinks are generally restricted to just a few locations: at the wing veins, along the edges of the wing, and at points or lines along the midpoint of the 'wing cells', the compartments defined by the vein network. Moreover, whereas Kühn and von Engelhardt assumed that their 'determination waves' issued across the whole wing, it is now clear that each wing cell has its own autonomous set of morphogen sources and sinks. So explaining the wing pattern as a whole can be reduced to the rather simpler problem of explaining the pattern in each wing cell. These patterns are typically arrangements of basic elements such as stripes and eyespots (ocelli), which in turn are thought to be induced by sources and sinks of morphogens called organizing centres. The morphogens diffuse through the wing cell, throwing biochemical switches where they surpass some concentration threshold.

How do these sources and sinks create the vast array of patterns that we see in nature? Nijhout has shown that simple combinations of them, located at wing tips, midpoints or veins, suffice to give a seemingly endless variety of pattern features. He suggests that butterflies and moths have at their disposal a basic toolbox of pattern-forming elements—just as Nabokov suspected (Fig. 4.27a). These define the concentration contours of the morphogens. As any of these contours can in principle represent the threshold above which the patterning switch is thrown, a single set of 'tools' can generate a wide range of structures (Fig. 4.27b). Not all these possible structures have been found so far in nature, so butterflies might not make use of the full 'morphospace' of patterns available to them. Perhaps there are some patterns that are detrimental to evolutionary success.*

---

*Again, it is not clear that all patterns have to be *useful* in evolutionary terms: as we have seen, some biological forms and patterns may be *neutral*, without adaptive value. This concept of neutral evolution was first hinted at by the nineteenth-century embryologist Karl Ernst von Baer.

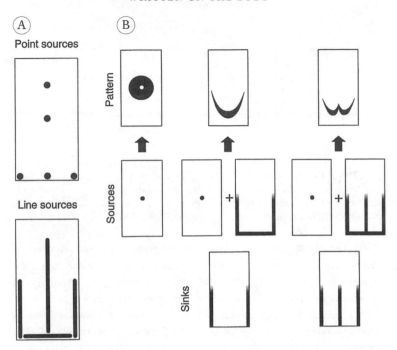

FIG. 4.27 A set of sources and sinks of morphogen in an idealized wing (*a*, here shown as a rectangular unit with veins at the edges and the wing edge along the bottom) can be combined to generate many of the pattern features observed in nature (*b*). (After Nijhout, 1991.)

Sources and sinks of morphogens must arise from some process that breaks the uniformity—the symmetry—of the wing cell. At the outset, the only 'special' places in the wing cell are the edges, at the veins and the wing tips. But some of the elements of the patterning toolkit are situated elsewhere in the wing cell. Nijhout has shown that such features can be produced by an activator–inhibitor scheme in which an activator diffuses from the vein edges into an initially uniform mixture of activator and inhibitor. At first, this leads to inhibition of activator production adjacent to the veins. Then a region of enhanced activator production appears down the wing cell midpoint (Fig. 4.28). This retracts towards the wing cell edge, leaving one or more point sources of activator as it goes. These sources could serve as the centres of eyespots, for example.

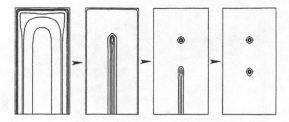

FIG. 4.28   The elements of the toolbox in Fig. 4.27a can be produced from an activator–inhibitor model in which an activator is released from the wing veins. The pattern of activator production (shown as contours) changes over time to become a central line that retracts to leave one or more spots. (Images: H. Frederik Nijhout.)

To verify this picture of butterfly patterning, we would need to identify and to monitor the putative morphogens. Ultimately these are thought to be produced by the activation of specific genes. Many genes have been identified that control pattern features in butterflies and moths, for example by changing colours, adding or removing elements, or changing their size. But how the genes exert these effects—presumably via diffusing morphogens—is in general still poorly understood. One of the best-studied pattern features is the eyespot or ocellus, a roughly circular target pattern (Fig. 4.29). These markings often seem to act as a defence mechanism, startling would-be predators with their resemblance to the eyes of some larger and potentially dangerous creature. Experiments by Sean Carroll of the Howard Hughes Medical Institute in Wisconsin and his colleagues have uncovered the genetic basis of the eyespot patterning process.

They found that a gene called *Distal-less* determines where the eyespots form. The gene is turned on (in other words, the Distal-less protein encoded by the *Distal-less* gene is expressed*) in the late stages of larval growth, while the butterfly is still in its cocoon. That the *Distal-less* gene is involved in this process was something of a surprise, since in arthropods such as beetles it was known to have a completely different role,

---

*The names of genes are conventionally spelt in italics, while the protein products derived from them have the same name in normal typeface.

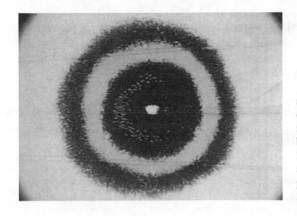

FIG. 4.29 The eyespot pattern is found on many butterflies and moths. It probably serves to alarm potential predators. (Photo: H. Frederik Nijhout.)

determining where the legs grow. We will see in Chapter 7 why the gene has this strange versatility.

Expression of the Distal-less protein occurs initially in a broad region around the tip of the wing, and the protein spreads by diffusion. Gradually, the production of the Distal-less protein becomes focused into spots, which define the centres of the future eyespots (Fig. 4.30). This focusing is suggestively similar to that seen in Nijhout's model for the formation of point-like morphogen sources. Once the focal points have been defined, they serve as organizing centres for the formation of the concentric rings. It seems that the Distal-less protein orchestrates this process. It becomes expressed in an expanding circular field centred on the focal point, and this signal controls the developmental pathways of surrounding cells, gearing them to produce scales of a different colour from the background. This process of differentiation of scale-producing cells around the eyespot focus is still imperfectly understood, but it appears to involve a suite of genes that also crops up, in different roles, in the development of other organisms. For example, genes called *Spalt* and *Engrailed* are involved in making the rings around the central eyespot of *Distal-less*. The boundaries are generally rather sharp, and Carroll and his colleagues think that the smooth variations in concentrations of the proteins as they diffuse away from the eyespot focus are converted into sharp-edged concentric patterns by the kind of threshold effect I have already mentioned: a 'switch' is thrown once the concentration reaches or

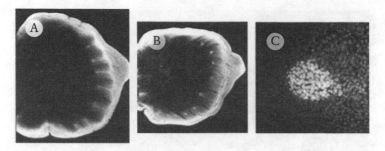

FIG. 4.30   The expression of the protein Distal-less in a caterpillar wing (seen here in light regions) marks out the centres of future eyespots. The distribution of the protein begins as narrow lines (a) that retract to leave single spots (b, shown close up in c). (Photos: Sean Carroll, University of Wisconsin-Madison, from Carroll, 2005.)

falls below a certain level. We'll see later that this is a common means by which diffusing biological morphogens establish abrupt boundaries between different tissues in a developing organism.

The toolbox of butterfly wing patterns seems able to fashion just about any pattern (Fig. 4.31a). But Turing's reaction–diffusion mechanism shows how few ingredients are needed to achieve such variety—indeed, the butterfly alphabet shown here is echoed in that generated by Andrzej Kawczynski and Bartlomiej Legawiec of the Polish Academy of Sciences in a computer model of Turing patterns that may form in a rectangular dish augmented by a few judiciously placed sources and sinks (Fig. 4.31b). We need not assume that mimicry in butterflies results from a laborious recapitulation, by the mimic species, of the evolutionary route that produced the pattern in the species being mimicked. If all butterflies share the same toolbox, then they have similar pattern repertoires, so that only a little tweaking is needed to copy the advantageous markings of another. This, Nabokov insisted, is how the viceroy butterfly mimics the monarch:* not by

---

*This was long assumed to be a classic example of Batesian mimicry. But it was shown in 1991 that the monarch and viceroy butterflies seem equally unpalatable to the birds that prey on them. It appears that their resemblance may instead be an example of so-called Müllerian mimicry, identified by the German zoologist Fritz Müller in 1878, in which two noxious species get the benefit of mutual enhancement of deterrence by copying each other's markings. Nabokov actually suspected as much, having put himself through the ordeal of tasting both species of butterfly and finding them equally bitter.

FIG. 4.31 Butterfly markings can be found that correspond to all the letters of the alphabet (a). A comparable 'alphabet' has also been spelled out from Turing structures created with strategically placed sources and sinks (b). (a, Copyright 2007 Kjell B. Sandved. b, from Kawczynski and Legawiec, 2001.)

gradual change, but by a small adjustment in the toolkit that produces a big change in pattern.

There is evidence that nature does make judicious use of the variations on a theme that the butterfly toolkit offers. For example, females of the African mocker swallowtail *Papilio dardanus* display at least 14 different patterns, known as morphs. Most of these are Batesian mimics of other, distasteful species—but some are not. All of the morphs, however, are produced by differences in a single gene. In other words, this tweak alone is enough to open up a great diversity of patterning. Nijhout has found that the mimicking morphs have patterns that are more prone to small variations from one individual to the next than are the non-mimicking morphs. That might seem the opposite of what we would expect—one might think that if you're going to be a mimic, you have more cause to get the pattern perfect than if your wing markings are there just for their own sake. But Nijhout suggests that so long as a morph does a passable job of mimickry, it is less vulnerable to predators than non-mimics. This means that the selective pressure that refines the way the butterfly looks is eased, and so there is more room for slight variations.

What case histories like this show is that, so far as biological pattern formation is concerned, the relationship between spontaneity and evolution is subtle. As D'Arcy Thompson (and Nabokov) suspected, nature makes patterns using universal processes that have their own characteristic forms. Turing identified one of these processes, which in its simplest manifestation offers a relatively limited range of options. While it looks likely that biology uses these 'off the shelf', it seems it is also capable of creating dazzling and almost limitless variations. But these all appear within an evolutionary context, which means that nature reserves the right to pick and choose patterns that 'work'. Living nature is inherently creative, but she is also capable of assessing her creations.

# RHYTHMS OF THE WILD

## Crystal Communities

When Alfred Lotka concocted his model of oscillating communities of predators and prey in 1910, he was by no means the first to consider what governs the proportions of a population. But that was a question more commonly considered by economists and political philosophers—those who studied the ebb and flow of human society—than by scientists. Apart from the effects of bacteria, viruses, and themselves, humans have not suffered much from predators or parasites for thousands of years. As a result, their numbers have risen fairly steadily, which could be seen even in the rudimentary social statistics of the eighteenth century. It was then that the Scottish philosopher Robert Wallace published (anonymously) *A Dissertation on the Numbers of Mankind in Ancient and Modern Times* (1753), which argued that in principle mankind should double its number every generation or so, in which case the global population grows by so-called geometrical progression: twofold, then fourfold, and then eightfold, and so on. But history does not bear this out, Wallace said, for such a rapid increase is prevented by various natural causes.

Wallace's book was read by the English clergyman Thomas Robert Malthus, an economist who was influenced by Adam Smith's notion of 'natural laws' that governed human society. Malthus drew together ideas that he found in Wallace and Smith, and also in the works of the Scottish philosopher David Hume, to explain why unfettered geometric growth of a population was not possible. The means by which people sustain themselves cannot be increased at a sufficient pace to keep up with this growth in numbers, Malthus said: subsistence cannot grow faster than arithmetically, which is to say, by the addition of a constant amount from one year to the next. What this means is that population growth, plotted on a graph, is a curve that slopes upwards ever more steeply, while increases in subsistence trace out a straight line. Sooner or later the lines will cross: demand will outstrip supply, and then it is every man for himself.

When Malthus published his argument in a short book in 1798 titled *Essay on the Principle of Population*, he wasn't saying anything particularly new. But the work crystallized various strands of thought in political economics, and the *Essay*, enlarged by Malthus in 1803, had a tremendous impact. Karl Marx read it and discerned within it a prescription for social unrest and violent revolution. To the nineteenth-century Belgian astronomer and 'social physicist' Adolphe Quetelet it implied, much as Malthus had intended, that we cannot expect population growth to follow a geometric growth curve, but that instead it traces out an S-shaped curve on which the size of a population, checked by scarcity of resources, levels off in a plateau. Quetelet's colleague Pierre Verhulst calculated the mathematical form of this curve, which he called the logistic curve. But Malthus was also read by the young naturalist Charles Darwin, who drew from it the message that life is a competition for limited resources, creating the selective pressure that drives evolution.

Darwin's theory provides a beautiful explanation for why the living world is so diverse and so rich. Species evolve, he said, to exploit the variety of conditions and resources in the environment—to occupy *niches*. Moisture and fertile soil supports fruit trees; insects live on the fruits; birds live on the insects and the fruits; larger birds live on smaller birds. Bacteria thrive in and among just about everything. Thus any habitable region can evolve a complex hierarchy of niches, with organisms adapted to each of them: they become ecosystems in which each species creates the conditions for the others. Some ecosystems support only a few dominant species; in others, millions of species coexist in a square metre of ground.

The heterogeneity of ecosystems may reflect that of the environment: marine organisms become adapted to fine gradations in water temperature, and when ocean currents or climate change, this may promote the colonization of a region by a species from outside. But the geographical diversity of biological populations cannot be explained by traditional Darwinism alone. For even apparently uniform environments rarely show uniform distributions of the organisms they sustain: the populations may be fragmented and patchy, and these patches might shift over time. There is, in other words, a kind of ecological symmetry breaking, and in some cases a degree of spatial patterning that is remarkable in its orderliness.

There are many reasons why living communities develop patterns, some of which I shall discuss elsewhere in this series. Here I want to explain why some of these can be regarded as precisely the same kind of processes as those that send ripples through chemical mixtures and that seemingly create the camouflage patterns of the zebra and the leopard.

## CYCLING TO SAFETY

Malthus considered that humanity was rather far from approaching the saturation point at which population growth exceeds resources. Verhulst was less sure: Belgium, he said, could expect to support no more than eight million people (the population was a little over half that when he wrote, but is around ten million today). The American population ecologist Raymond Pearl rediscovered Verhulst's work on the logistic curve in the 1920s, when he claimed to have observed this pattern of growth in the United States population. Pearl, I might point out, was a great supporter of the mathematical approach to population dynamics developed by Lotka.

But at face value, Lotka's model seems to be saying something quite different from the predictions of S-shaped population growth—for the former says that populations fluctuate periodically in their size, the latter that they reach a plateau and stay there. Yet Vito Volterra, the Italian zoologist who studied Lotka's model in the context of fish populations in the Adriatic in the 1920s (page 116), showed that there is no inconsistency. He found that Lotka's equations predict that both situations are possible: the population might settle into a steady state, or oscillate indefinitely. The former case is unlikely, however, because it takes only a small disturbance to set off the wobbles. In the language introduced in Chapter 3, the Lotka–Volterra model (as it is now known) is apt to undergo to a Hopf bifurcation.

We have seen above why this happens: in a system of predators and prey, populations tend to overshoot. Given abundant prey, the predators gorge until they find that food has become scarce, in which case their numbers crash soon after, favouring re-expansion of the prey population. The fine-tuning needed to achieve stable populations is rarely attained; more commonly, as the popularizing Darwinist Herbert Spencer pointed out,

> Every species of plant and animal is perpetually undergoing a rhythmical variation in number—now from abundance of food and absence of enemies rising above its average, and then by a consequent scarcity of food and abundance of enemies being depressed below its average.

Herbivorous animals commonly show such cycles: those of voles and lemmings last around four to six years, while larger creatures such as muskrats have nine- to ten-year cycles. Predators that prey on these creatures often tend to have cyclic variations with the same period, and it is tempting to see these as the result of interactions of the Lotka–Volterra type, which lock the two cycles together. But the reality is almost certainly not so simple. Take, for instance, the interactions between snowshoe hares and the lynxes that prey on them in eastern Canada. This is one of the few systems for which there are long records, because both animals have long been captured for their pelts by trappers for the Hudson Bay Company. Records of the number of fur catches have been kept since 1845, and if one assumes that the trappers always catch a fixed proportion of the population, the ups and downs of the catches should reflect those of the populations as a whole.

The numbers of both lynx and hare catches oscillate with almost a ten-year cycle (Fig. 5.1), with the two oscillations slightly out of step, apparently as predicted by the Lotka–Volterra scheme. But if we look closely at the records, we can see that sometimes the predator cycles precede those of the prey, implying that the hares are eating the lynxes! What is more, the life cycle of the lynx is such that its population grows considerably

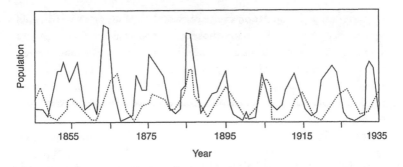

FIG. 5.1 Roughly periodic fluctuations in population size are evident in the records of lynx and snowshoe hare catches kept by the Hudson Bay Company since the mid-nineteenth century. Both the predator (lynx: dashed line) and the prey (hare: solid line) populations oscillate with a period of about ten years. Are these Lotka–Volterra cycles?

more slowly than that of the hares, and under these conditions Lotka–Volterra cycles are not expected because the predators cannot expand fast enough to overtake and control the prey population.

Thus the Lotka–Volterra model does not explain everything about ecosystem oscillations. Nevertheless, it seems likely that the basic principles it embodies—growth limited by resources or by predation, and interactions that lead to oscillatory instabilities—are responsible for the periodic variations seen in many real populations. In fact, evolution has apparently acted in some cases to turn population oscillations to adaptive advantage. Cicadas of the genus *Magicicada* spend most of their time underground, but emerge every 13 or 17 years to mate and then to die in a matter of weeks. These are the longest life cycles known for any insect, and the so-called 'periodic cicadas' are also unusual in that their cycles are synchronized within a particular colony—they all emerge at the same time.

These are genetically hard-wired cycles of birth and death, whereas the population cycles of mammals are simply changes in numbers that bear no relation to the creatures' life span. But why 13 and 17? These are both prime numbers, having no divisors (other than 1) smaller than themselves, and Mario Markus of the Max Planck Institute for Molecular Physiology in Dortmund and his co-workers have argued that this is no coincidence. If the cicadas were to emerge every 12 years rather than 13, any predators (or parasites) with population cycles of two, three, four or six years could synchronize these cycles with those of the cicadas so that they are abundant every time the insects emerge. For the prime-number cycles, in contrast, if the life cycle of the predators is any number $n$ (less than 13 or 17) then there can be an exact coincidence between cicada emergence and the peak of predator abundance only once every $13 \times n$ or $17 \times n$ years— so this optimal situation for predators is very rare. Prime numbers make it impossible for predators with shorter life cycles to get in step more often than this.

Markus and colleagues devised a mathematical model in which predators and prey could randomly alter their life cycles by mutation. The prey that successfully evade abundant predators are 'fitter' in evolutionary terms, and as the model plays out over time, these fitter prey will come to predominate in the population. The researchers found that the model produced prey populations with prime-number life cycles.

The only snag in this neat argument is that there are no known 'cyclic' predators of the cicada (its main predators are birds). But it is possible that the prime-number cycles were set in place at an earlier stage of evolutionary development, when cicadas were stalked by creatures that have since become extinct—unable, perhaps, to keep up with the insects' avoidance strategy. In other words, the absence of periodic predators may in itself be an indication of the success of the prime-number strategy.

For many years it was assumed that the fundamental character of simple predator–prey interactions is oscillatory. But in the mid-1970s ecologists began to realize that even simple models of population dynamics display more richness and complexity than the periodic oscillations of the Lotka–Volterra scheme. For example, if the rates of population growth and predation success are changed so that the predator and prey populations become more sensitive to fluctuations in each other's numbers, these oscillations may undergo period-doubling bifurcations such as those that we observed previously in the BZ reaction. This can lead to increasingly complex periodic oscillations, and ultimately to chaos—to fluctuations in population density without any apparent regularity at all.

You do not even need a predator to destabilize a population and tip it into oscillatory cycles: overcrowding alone will do the job. That even the simplest of population models can show dramatic and unpredictable ups and downs was demonstrated in the 1970s by Robert May, then at Princeton University, and George Oster at the University of California at Berkeley. They took a close look at a deceptively simple mathematical model of a population that breeds seasonally to produce generations that do not overlap. Many insect populations are like this. In this model the size of each generation grows in proportion to that of the previous generation when the sizes are small, but is inhibited by overcrowding when the size approaches some critical threshold. This means that the model is *nonlinear*—cause (the size of the preceding generation) and effect (the size of the ensuing one) are not related in direct (linear) proportion to one another. Nonlinearities are a nearly ubiquitous cause of complex behaviour and pattern formation; the positive and negative feedbacks exhibited in BZ and Turing processes are also examples of nonlinear behaviour.

The behaviour of this model, called the logistic model, depends in part on how sensitive the size of each successive population is to the size of the previous one. If this sensitivity is low, the population simply dies out: the reproductive success is not high enough. For intermediate sensitivity, the population settles down to a steady value—growth never becomes so great that it overwhelms the resources. But when if the population grows at too great a rate, things get complicated: the population oscillates in size between greater and smaller values in successive generations. For even larger growth rates there is a series of period-doubling bifurcations, so that the cycles repeat every two oscillations, every four, every eight and so on (Fig. 5.2). For a sufficiently large sensitivity, the fluctuations appear to be irregular; in fact, they are chaotic. This means that, although the equations of the logistic model specify the population size exactly—there are no random elements in the model—in practice it becomes impossible to predict how big the population will be at any moment. This is because the state of a chaotic system is acutely sensitive to barely perceptible differences in its initial conditions.

Ecologists had long known that real populations undergo irregular fluctuations in size. But they had assumed that these were the result of the unpredictable influences of the environment—changes in the weather, in crop yields, and so on. Such influences undoubtedly play a role in introducing randomness (which physicists call noise) to population

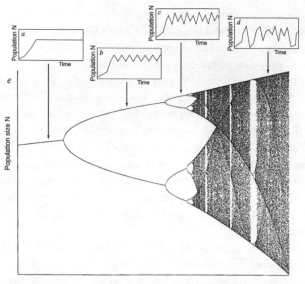

FIG. 5.2 A simple model of a population that grows exponentially until limited by overcrowding generates complex behaviour. For different values of the 'sensitivity parameter', the model can produce a uniform population size (*a*), simple oscillations (*b*), period-doubled oscillations (*c*), or more complex oscillations that lead eventually to irregular, chaotic fluctuations (*d*). A plot of the population sizes against the sensitivity parameter (*e*) shows this sequence of period-doubling bifurcations: a cascade that gets increasingly more intricate until eventually descending into chaos, corresponding here to a dense 'dust' of data points on the right.

dynamics, but what May and Oster showed was that there can also be an intrinsic chaotic unpredictability in the size of populations, irrespective of external factors.

One of the consequences of this behaviour is that the effects of perturbations to a population can be hard to predict. This, in fact, is what prompted Volterra to study the problem in the first place: he was attempting to understand why the relative proportions of predators and prey in fishermen's catches altered when the intensity of fishing activity altered (during wartime, for instance). The simplistic expectation would be that, while less fishing would deplete the fish stocks less rapidly and so lead to greater numbers in each catch, the effect would be the same for predators and prey. But in a nonlinear system that need not be so, since the effects on future populations of a sudden decrease in both predators and prey (because of intense fishing, say) depend on the relative rates with which the two populations recover from this change, and will not generally be identical.

## CLAIMING YOUR PATCH

Volterra's analysis assumes that the predator and prey populations are always uniform in space, like a well-mixed Belousov–Zhabotinsky mixture. But real populations are usually patchy: at any instant, there is likely to be some clustering of creatures separated by more sparsely populated regions. Partly this might be a result of geographical features such as vegetation differences that will influence the way in which the creatures distribute themselves over the terrain. But such fluctuations in population density may also arise by chance, simply because the ecosystem is not 'well stirred'.

Yet as the BZ reaction indicates, small variations in density can have significant consequences when they occur in nonlinear systems. The fluctuations can be amplified by positive feedback, and local irregularities can propagate to distant regions by triggering travelling waves. Once we recognize that this propagation must happen on a timescale determined by the rates at which the animals move from place to place, then the Lotka–Volterra scheme becomes entirely equivalent to a reaction–diffusion mechanism, raising the possibility of spatial patterns in the numbers of predators and prey. In particular, in the idealized ecosystem of rabbits and foxes (page 121), the rabbits can be regarded as an activator species that multiplies itself locally, while the foxes are inhibitors that act over long ranges. That should lead us also to anticipate stationary Turing patterns.

The possibility of travelling waves and spatial patterning in predator–prey models has been appreciated ever since the work of biologist Ronald Fisher in the 1930s, though it is only in recent years that the idea has been given much attention. While field biologists who measure the densities of populations in their natural habitat have long found that these can vary tremendously from place to place, such variations were again assumed to be the result of noise (random structure or conditions) in the environment, and were therefore regarded as a nuisance that simply obscures the underlying 'true' dispersal behaviour of populations. Talking about parasites and their hosts, a specific subclass of predator–prey systems, the biologist Peter Kareiva from the University of Washington said in 1990 that 'as recently as a decade ago, any field ecologist who recorded widely scattered rates of parasitism bearing no relationship to the density of hosts would probably have shelved the data as useless'. But, says Kareiva, 'we now know that such "disorder" can be a source of "order" in species interactions'. What he meant by this is that apparently random patchiness in populations can in fact be necessary to allow predators and prey to coexist in a stable state. This patchiness may be an intrinsic outcome of the interaction, not a result of superimposed noise.

Working at the Santa Fe Institute in New Mexico, physicist Ricard Solé and his co-workers found that patchiness can appear in a simple population model in which two species compete for territory. This is a more general picture of Darwinian competition than that offered by predator–prey models: the two species do not necessarily interact directly at all, but simply try to outdo each other in foraging for food or other resources. Classical theories of ecology predict two possible outcomes. If the two species are not in strong competition—if their growth rates are slow, for instance, or the resources are abundant—then they can coexist in the same region. But if the competition becomes more intense, there is 'competitive exclusion', otherwise known through the phrase 'this town ain't big enough for the both of us': one of the species is eliminated by the other.

In the early 1990s, Solé and his co-workers investigated such a model in a computer simulation. They scattered two competing species randomly over a patch of ground and watched to see how the populations evolved according to the rules that governed the competition between them. The model predicted that an initially uniform distribution breaks up—that is, its symmetry is broken—into patches where one or other of the species dominates. Symmetry breaking permits the species to coexist where uniformity would not.

This is seen most clearly in work conducted at the same time by Michael Hassell and Hugh Comins of Imperial College in London in collaboration with Robert May. They studied spatial variations in populations of parasitoids and their hosts. Parasitoids are a particularly nasty kind of parasite: they are insects that lay their eggs in (or close to) the host's body so that the parasitoid larvae may devour and kill the host once they hatch. Here parasitoids are equivalent to predators, and their hosts are the prey.

Host–parasitoid interactions in nature can display periodic oscillations very much like those seen in other predator–prey systems, with the host and parasitoid populations rising and falling with the same period but out of step with one another. With some subtle differences from true predation, the interaction can be described by a mathematical model similar to the Lotka–Volterra scheme. But when the two populations are assumed to be distributed uniformly across a landscape, such a model generates oscillations of ever increasing amplitude. This is an unstable outcome, which means that eventually the hosts and then the parasitoids that feed on them are driven to extinction. The implication is that, according to this model, the host and parasitoid populations cannot coexist with each other in the long term.

But what if the parasitoids are distributed in an uneven manner? Hassell and colleagues showed that if the variability in the density of parasitoids searching for hosts was great enough, the two populations could both manage to persist indefinitely. For a sufficiently variable parasitoid distribution, there would always be regions in which the hosts would escape predation. Within this picture, the spatial patterning (patchiness) seen in nature is not just noise scattered over the underlying population dynamics but an essential stabilizing factor, without which the ecosystem would collapse.

What form might this patchiness take? The researchers studied the spatial structure of their ecosystem in a computer model in which the environment was depicted as a square grid of cells, each of which contained a certain number of parasitoid and host organisms. An initially random distribution of these species evolved by spreading from cell to cell in what is basically a form of diffusion. The host and parasitoids in each cell 'react' with one another according to mathematical equations that describe the reproduction of both populations and the killing of hosts by parasitoids. In other words, this model again has the ingredients of a reaction–diffusion process.

This cellular model produces a variety of spatial patterns that depend on the rate at which the organisms spread from cell to cell. For a certain range of spreading rates, dynamic spiral waves of population density appear (Fig. 5.3a). For other spreading rates, the researchers found chaotic and constantly shifting patterns instead (Fig. 5.3b). But if the parasitoids disperse much more quickly than the hosts, the populations can 'freeze'

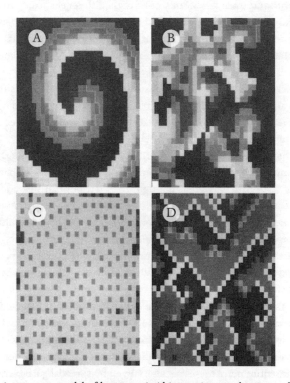

FIG. 5.3  A computer model of host–parasitoid interactions produces complex spatial patterns from patchy initial distributions. These include spiral waves (a), chaotic patterns (b, d), and static, almost regularly spaced islands of prey among a 'sea' of predators (c). The grey scale here shows different relative abundances of host and parasitoid in each patch. (Images: Michael Hassell, Imperial College, London.)

into a crystal-like lattice of small patches spaced at roughly regular intervals and containing high densities of hosts, surrounded by wide regions heavily populated with parasitoids (Fig. 5.3c). These are comparable to Turing structures generated by long-ranged inhibition.

The point about all of these patterns is that they represent stable states—even though the chaotic patterns and the travelling waves change constantly, the populations never collapse as they do if either they are uniformly distributed or if cell-to-cell migration (diffusion) is prohibited. So, again, the spatial patterning here has the non-trivial ecological consequence of allowing otherwise unstable predator communities and their prey to survive in the same environment through a game of predatory hide and seek. This behaviour seems to be borne out by laboratory studies of predatory mites and their prey conducted by biologist Carl Huffaker of the University of California in 1958. He found that by imposing patchiness on the mite populations and maintaining it by restricting the mites' freedom to move around (which involved creating a maze of Vaseline barriers weaving amongst the food), the predators and prey could coexist for almost seven times as long as they could if the mites were unobstructed and the patchiness was therefore liable to become smoothed out.

There is at least one important lesson in these discoveries for our attempts to manage wildlife habitats and ecosystems: space matters. Some ecosystems need space to spread, so that they can organize themselves into patchy communities that coexist where uniform ones cannot. The more we carve up the environment into isolated parcels by building roads and other barriers to the dispersal of species, the more we destroy the capacity of populations and ecosystems to use spatial patterning as a means of survival. What is more, we would be foolish to assume that, in a complex ecosystem like this, stability and survival depend in a simple way on habitat fragmentation. Solé and his colleague Jordi Bascompte have found that, once a community cannot spread in a connected domain across an entire territory—that is, once an individual organism cannot pass from one edge to another without stepping outside of a populated domain—the average patch size shrinks rapidly and the species struggles to survive.

## THE QUICK AND THE DEAD

That anything so organized as a spiral wave could arise spontaneously in natural populations is too much for many ecologists to accept, and these pattern-forming models have been dismissed by some as little more than a means of generating pretty patterns on a computer. One criticism that has been raised is that the models are too perfect; in real ecosystems there are randomizing elements, such as variations in landscape features and vegetation, which might wash out any elegant patterns such as spirals. But researchers have shown that spiral patterns can persist even in the face of such 'noise'. Indeed, noise may do more than just generate a degree of fuzziness—it might be intrinsic to the patterning in time and space. Kevin Higgins of the University of California at Davis and colleagues have shown that the large fluctuations in population size of the Dungeness crab off the North American west coast, which show some indication of a ten-year periodicity, can be reproduced by a mathematical model of the ecosystem dynamics only when small but significant random environmental perturbations are injected into the equations. Without the noise, the same model predicts a stable population size, which is quite different from what is observed.

Similarly, Ricard Solé and his co-workers José Vilar and Miguel Rubí have suggested that noise is essential to account for the spatial patterns in some predator–prey communities. They have looked at the patchy distribution of plankton in the sea. Here the predators are the zooplankton, microscopic animals whose prey are the phytoplankton, the sea's tiny plant life. The zooplankton can swim around and so have a faster diffusion rate than their prey, which are simply carried passively by the ocean currents. Both predator and prey are distributed in a patchy fashion, but the phytoplankton patches are bigger than the zooplankton patches. Activator–inhibitor models of this ecosystem, based on Lotka–Volterra equations, give rise to blotchy Turing-type distributions of the two communities—but with precisely the reverse of the observed characteristics, with the predator patches being larger. Solé and colleagues showed that by adding a random, 'noisy' aspect to the distribution of predators in the activator–inhibitor scheme they could obtain just the kind of patchiness found in reality. 'Noise not only generates patterns but is also able to produce the right ones', the researchers say.

Although they may be robust against noise, spiral waves appear only for rather specific conditions in the cellular model of Hassell and colleagues. If the parasitoids disperse to neighbouring cells more rapidly than the hosts (which is likely in practice), then disordered, chaotic patterns are favoured instead. So it might be very hard to find ecological spiral waves in nature. But looking for chaotic patterns could be more fruitful. What does such a search reveal?

These ideas of spontaneous patterning are still new enough in population biology that there have been few serious attempts to verify them in field studies. Although we know that natural populations tend to be distributed highly unevenly, it is no mean feat to distinguish chaotic 'intrinsic' patterns from the variability imposed by external sources of randomness. Were the patterns seen in Carl Huffaker's experiments of the former or the latter sort, for instance? No one has checked.

But in 2002 a team of European scientists found evidence of genuine Turing-pattern clustering in ant colonies. Guy Theraulaz of the Université Paul Sabatier in Toulouse and his co-workers raised colonies of *Messor sancta* ants in Petri dishes. These insects have the remarkable trait of collecting the dead bodies of colony members and arranging them in piles, or ant cemeteries. The researchers scattered dead ant bodies evenly around the circular perimeter of the dish, where live ants entering the 'arena' tended to stay. As the corpses were added to the dish, the live ants began picking up bodies and organizing them into piles.

This disposal process does not seem to be done particularly efficiently. Instead of choosing a few locations for cemeteries and sticking to them, the ants seemed at first to start piles of bodies at random, some of which would simply be dismantled as ants subsequently carted the bodies away. Over the course of three hours, the number of cemeteries rose steeply until it reached a maximum, after which this number decreased slightly and attained a roughly steady value. By this point, particular cemeteries had became well established, so that, although the ants continued to pick up and drop bodies, the spatial pattern of the piles stayed constant overall (Fig. 5.4).

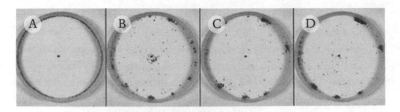

FIG. 5.4   Ant cemeteries (piles of dead bodies) that emerge as ants redistribute corpses placed uniformly (a) around the perimeter of a circular dish. Here I show the pattern's evolution after 6 (b), 12 (c) and 45 (d) hours. (Photos: Eric Bonabeau, Santa Fe Institute, New Mexico, from Theraulaz et al., 2002. Copyright 2002 National Academy of Science.)

These stable cemeteries are analogous to the spots of Turing patterns, although here the 'spots' are expressed in effectively just one dimension: a line along the edge of the arena. Although the components of the pattern—the ant corpses—continue to be moved around (analogous to the diffusion of molecules), the pattern itself becomes fixed. Theraulaz and colleagues showed that the changes in the number of cemeteries over time, and the average distance between them, could be explained by a model in which there is local activation and long-range inhibition in the formation of corpse piles. Local activation arises because ants are more likely to drop a body on a pile, and less likely to pick up a body from a pile, as the pile gets larger: the growth of piles is autocatalytic. Long-range inhibition stems from the fact that the region close to a big pile gets swept clear of bodies, and so it is less likely that a new pile will be started close to an existing one.

The researchers suspect that mechanisms like this might underlie many other aspects of habitat formation and grouping, such as nest construction, in higher organisms. Many animal and plant species gather together in groups, a distribution that is called positively contagious. Fish and plankton, for example, form schools (about which I have more to say in Book II). Animals that are strongly territorial, on the other hand, such as songbirds, will distribute themselves so as to stay an optimal distance from their neighbours, a situation said to be negatively contagious. In the former case it is as if each individual in the community is attractive to others, and in the latter case as if they are repulsive. For some species, there is an ideal balance of attraction and repulsion: there is safety in numbers (the predator is less likely to pick you if you're one of a crowd) but too great a number incurs a risk of overcrowding and depletion of local resources. In striking a balance between these two factors, creatures may find themselves aggregating into small clusters that are more or less periodically spaced. This is seen, for example, in the pattern of nesting territories of certain fish. Male Mozambique tilapia and bluegill sunfish both dig pits into the sandy bed of lakes to define their territory. An interplay of attraction and repulsion results in each male fish staking its claim roughly an equal distance from the others, which means that, on average, the centres of the pits tend to be arranged on a honeycomb lattice, with the ridges of the pits defining a hexagonal network (Fig. 5.5). The same effect drives royal terns on the south-eastern American coast to arrange their nests in an approximately hexagonal array, the rims traced out emphatically in faeces.

Animals can diffuse; but plants have to stay put. Vegetation can be patchy, but you might imagine that this is down to mere chance, dependent on where seeds happen to fall. That, however, should give rise to patches of all sizes, with no characteristic distance between them. Yet some vegetation shows the kind of non-random patchy distribution you

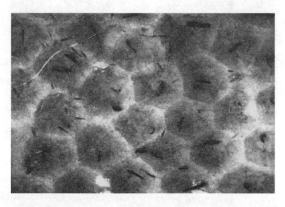

FIG. 5.5   Honeycomb nest territories of male *Tilapia mossambica* fish. (Photo: courtesy of Princeton University Press; from Barlow, 1974.)

might now recognize as symptomatic of pattern-forming systems, in which the features have a characteristic size and separation. Alpine shrubs colonize uniform ground in clumps of roughly the same size, for example, while plants that need a lot of root space distribute themselves so as to stay an optimal distance from their neighbours: they too are negatively contagious. Vegetation patterns seen in semi-arid and desert regions in Africa, the Middle East, Australia, and elsewhere exemplify this sort of distribution, consisting of spots and stripes that immediately make us think of animal pelts (Fig. 5.6)—and suggesting that some reaction–diffusion process might be at play here too.

That is what Christopher Klausmeier of the University of Minnesota proposed in 1999: he thinks that such clumps of vegetation might be related to Turing patterns. The growth of vegetation depends on scant water resources, and in particular on the plant's ability to capture rain water and prevent it from running away down a hill slope. A clump of grass then becomes a local activator of more growth because it blocks flowing water and stops it from running off. But in doing so, it deprives the ground further downhill of water, and so creates longer-ranged inhibition. Klausmeier's mathematical model of this process generated

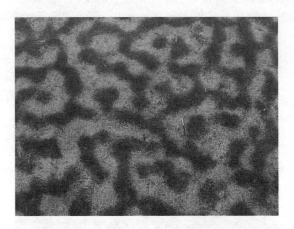

FIG. 5.6 A labyrinth patterns of perennial grass in the northern Negev desert. The average distance between patches is about 15 cm. (Photo: Ehud Meron, Ben Gurion University. From von Hardenberg *et al.*, 2001.)

stripes of vegetation on sloping ground which move slowly uphill—they are not exactly Turing patterns, but travelling waves that follow the hill contours. On flat ground the patchiness he predicted was more random and spotty, and reflected small variations in the underlying height of the ground: in this case the feedback effects inherent in the model amplified these seemingly insignificant variations and turned them into a patchy pattern.

Researchers at Ben Gurion University in Israel have looked at how the patterns generated by this kind of model are affected by variations in rainfall. On flat ground they saw that the vegetation patches changed from spots to stripes to carpets punctured by holes as the amount of rainfall increased. All these patterns are seen in nature (Fig. 5.7). Sometimes the researchers found that spots grew into rings, as intense competition for water left a vegetated patch depleted in the centre. They also found that the changes that took place as rainfall increased were not simply the reverse of those caused by a decrease: the amount of rain needed to trigger spots of vegetation on bare ground was greater than that at which partly covered ground would lose all vegetation due to declining rainfall. In other words, if drought kills off the vegetation, it is harder to get it growing again. So here the pattern depends on the past history of the system.

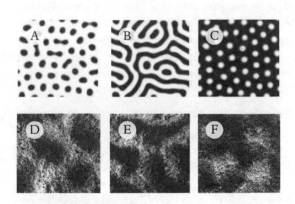

FIG. 5.7 Patterns of vegetation produced in a Turing-type model (a–c), and corresponding patterns seen in the Negev (d–f). (Photos: Ehud Meron, Ben Gurion University. From von Hardenberg et al., 2001.)

## TURING'S CATHEDRALS

Social insects make themselves the most marvellous homes. Some wasps create multi-storey combs of papery material from plant fibres mashed by their mandibles, the floors interconnected by passageways and the whole reaching to a metre or more in height (Fig. 5.8). Termites have been said to construct veritable 'cathedrals of clay'—*Macrotermes* mounds are conical spires that can soar to six or seven metres, laced with the most intricate networks of tunnels and chambers (Fig. 5.9) that include nurseries, air vents, 'gardens' for growing fungal food, and a royal palace. If the termites were scaled up to human size, their cities would tower to about a thousand metres.

It is tempting to call these insects master architects. But that is to give the wrong impression, for an architect works from a blueprint. Early efforts to explain the dwellings of social insects proposed that the creatures each carry such a blueprint in their heads, somehow possessing a spatial picture of the whole enterprise which they then proceed to construct piece by piece. This would be remarkable almost to the point of absurdity, and while we have seen that one should not underestimate the capabilities of insects, there has never been any evidence that they build their palaces this way.

Instead, these are structures that arise spontaneously from the interactions between individuals in the community, just like the Turing cemeteries of ants. They are self-made patterns of the same general sort as those of reaction–diffusion chemical systems. This is no mere metaphor: the analogy is precise, stemming from the same combination of autocatalytic

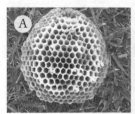 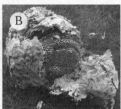 

FIG. 5.8 Wasp nests are typically more varied than honeybee combs. (Photos: *a*, Daniel R. Blume; *b*, Myrmi; *c*, Matt Foster.)

FIG. 5.9  The conical mound of a nest of *Macrotermes michaelseni* termites (*a*), and the internal structure as revealed by a cast (*b*). The nest consists of an elaborate labyrinth of surface conduits and deep ventilation tunnels. (Photos: *a*, Scott Turner, State University of New York, Syracuse; *b*, Rupert Soar, Loughborough University.)

positive feedback and inhibitory negative feedback. Social insects combine these generic principles with the specific features of their environment to create some of the most elaborate and awesome patterns in nature.

Moreover, the signals that guide construction are often conveyed by diffusing chemicals, comparable to Turing's morphogens. Ants, for instance, communicate with one another by emitting pheromones, just as *Dictyostelium* cells release a chemo-attractant. In perhaps the simplest case, a single individual creates a pheromone field that acts as a template for the activity of other insects around it. The queen of a colony of the termites *Macrotermes subhyalinus*, obscenely bloated with eggs, is housed in a 'royal chamber' that the workers erect around her, directed by the concentration of pheromone she emits. A worker deposits a soil pellet in the growing walls of the chamber only if the concentration of pheromone exceeds a certain threshold, or falls within a particular window of concentration—in which case the walls map out a three-dimensional contour of pheromone level (Fig. 5.10).

But the royal chamber is typically more complex than a simple dome, and its construction follows a sequence that seems to demand an

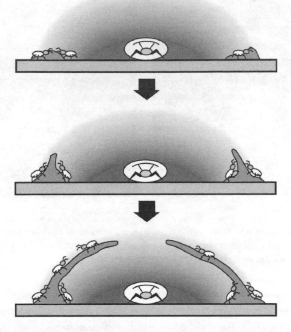

FIG. 5.10 How termites construct a 'royal chamber' around the template created by the emission of a pheromone (grey) from the egg-bloated queen.

impressive degree of cooperation among workers. The walls begin as pillars, which are first lengthened before being topped with a roof—made, if necessary, by workers standing on the queen's enormous abdomen. Finally, the pillars are linked up by building partition walls between them.

Where does this coordination come from? The fact is that the queen is not the only individual emitting a pheromone. The oral secretion used by workers to mix their soil pellets into a cement-like paste also contains a chemical attractant, which has a range of about 1–2 cm and acts for just a few minutes after being produced. This pheromone draws other workers to the source, encouraging them to add their pellets to the structure there.

That creates positive feedback: the more a pillar is augmented, the stronger a pheromone source it becomes. It is, in effect, a self-amplifying structure. Jean-Louis Deneubourg of the Free University of Brussels has shown that this process can be mimicked by a reaction–diffusion model, from which spots of preferential pellet deposition—incipient pillars—emerge spontaneously with roughly even spacings. In reality, pillars can also be initiated at pre-existing irregularities, such as objects lying on the ground, rather like a raindrop condensing around a speck of dust.

The building process in termite mounds appears to be guided by yet another chemical signal as well, for workers may emit a 'trail pheromone' that induces others to follow in their footsteps, recruiting them to the construction site close to the queen. Deneubourg found that if he added this element into his model, and made the assumption that the trail pheromone also discourages workers from depositing their pellets on the trail it marks out, the pillars are transformed into walls either side of the trails, creating galleries. There is some biological evidence that the trail pheromone does have this effect of preventing pellet deposition, which after all would make good sense: it is important that the trails do not become blocked.

In the convoluted tunnels of a real termite mound, air currents could influence the way pheromone signals are transmitted. Deneubourg and his co-workers have speculated that gyrating currents wafting along channels in the nests of the subterranean termite *Apicotermes arquieri* could play a role in the construction of the extraordinary helical ramps that give the insects access from one level to the next.

The key feature of these processes is that they require each insect to follow only local rules—to respond to the information it can gather about its immediate environment—and not to observe any global blueprint. These rules may establish amplifying and inhibitory feedbacks, but they differ from those of a chemical Turing system in that they can be considerably more complex. A termite registering a particular pheromone signal, for instance, could respond in a variety of ways, and those responses may vary over time. In the late 1950s the zoologist Pierre-Paul Grassé showed that the action a worker might take while building a termite nest depends on the structure of the nest so far. Once it reaches a certain stage, the mere presence of a particular structure could induce workers to begin a different building activity, which then in turn induces a further type of behaviour once it reaches maturity. Grassé called this process stigmergy. In this view, it could be said, the future is determined not by relentlessly following a single set of rules, but by responding to what has been done up to that point.

The question is, of course: what are the rules? One of the most thoroughly studied stigmergic nest-building activities of social insects is that of the wasps of the widespread genus *Polistes*, often called paper wasps because of the fibrous material from which they make their combs. Each nest is typically round and consists of 150 or so tubular cells with hexagonal cross-sections. One of the useful things about *Polistes* nest construction is that it can be tracked step by step in the laboratory: the wasps will use genuine paper to make it, if this is offered to them, and by giving them differently coloured paper at different stages one can create a colour-coded record of the progress of construction. In this way, researchers have found that the wasps seem to follow particular rules for adding new cells to the structure, rather than just adding them at random. For example, they will preferentially fill in 'gaps' at the edge of a layer, where there are three walls already built. Eric Bonabeau and his colleague Guy Theraulaz devised a model for investigating what kinds of structures might appear in such a process, governed by local rules that define the probability of a cell being added at particular types of site. They considered only two such rules, specifying the probabilities that a new cell would be added at sites with either two or three existing walls in place (Fig. 5.11). They also

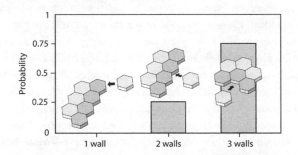

FIG. 5.11 Simple rules for building wasp nests. In this model for the nest structure of the wasp *Polistes dominulus*, new hexagonal cells are constructed with a higher probability if three walls are already built than if just two are built. There is zero probability of building a new cell onto an edge with just one wall.

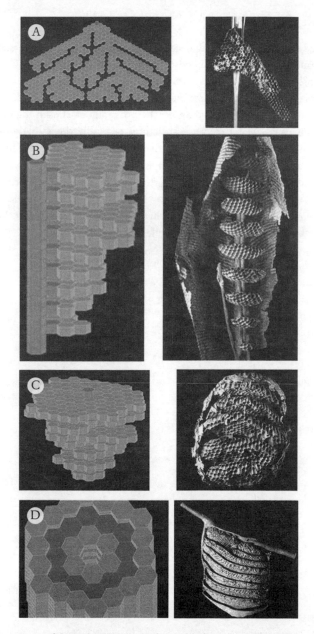

FIG. 5.12 Some of the nest architectures that result from the model. These resemble the nests built by various genuses of real wasps: *Agelaia* (a), *Parachartergus* (b), *Vespa* (c), *Chartergus* (d). (Images: Guy Theraulaz, Université Paul Sabatier, Toulouse.)

allowed for the possibility that cells might be stacked into layers. Using these rules, three-dimensional nest structures could be grown, rather like a tiling process.

These elements alone were enough for the model to generate a wide range of nest shapes, depending on the probabilities chosen for the two rules (Fig. 5.12). Many of these structures resemble those made by real wasps of different species, showing that all that is needed to produce this architectural diversity is a slight change in the propensities with which the wasps add new cells. The insects do not work to different blueprints, but merely have small differences in how they like to lay their tiles. When they begin, they have no more idea than we do of the pattern that the tiles will make. Here again, the pattern effectively makes itself.

# HOW DOES YOUR GARDEN GROW?

## The Mathematics of a Daisy

'From the contemplation of plants', wrote the seventeenth-century English botanist Nehemiah Grew, 'men might first be invited to Mathematical Enquirys.' D'Arcy Thompson thought that too, for he felt sure that the orderly arrangement of leaves or florets on a plant stem would have attracted the attention of the ancient Egyptian and Greek geometers. Aristotle's pupil Theophrastus commented on it, as did, much later, that princely observer of nature Leonardo da Vinci. But it was not until 1754 that the Swiss botanist Charles Bonnet explained clearly the fundamental feature of 'leaf ordering', known as *phyllotaxis*. The leaves, Bonnet pointed out, are disported around the stem in a spiral.

In 80 per cent of plant species, the succession of leaves up the stem traces out a spiral, with each leaf displaced above the one below by a more or less constant angle (Fig. 6.1a). The angle of offset is, in many different species, so often close to 137.5° that the fact demands explanation. Nearly all the remaining one-fifth of leafy plants show one or the other of just two alternative phyllotactic patterns. In one of these, called distichous, successive leaves sprout on opposite sides of the stem, usually with the leaf wrapped almost fully around the stem (Fig. 6.1b). We could regard this as a form of spiral too, in which the offset angle is 180°. The third pattern, called whorled, has little clusters (whorls) of leaves—two or more—at regular intervals up the stem, with each whorl offset so that it sits over the gaps of the whorl below. A common whorled pattern juxtaposes two leaves 180° apart offset at an angle of 90° from the two below (Fig. 6.1c)—this is called decussate. Mint has this structure, and so does the stinging nettle.

It is tempting to imagine that these arrangements are clever adaptations selected because they give the leaves maximum exposure to sunlight. We can be sure that arrangements that significantly obscured the sun's rays would be selected *against*; but there seems no clear evidence that the

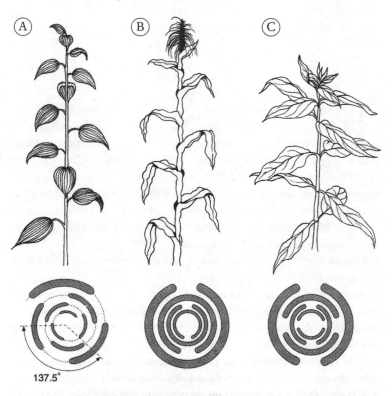

FIG. 6.1  Three distinct patterns can be identified in the arrangement of leaves around plant stems (phyllotaxis): spiral (*a*), distichous (*b*) and whorled (*c*). Below each drawing is a schematic representation of the leaf pattern as seen from above, with successive leaves depicted as smaller the farther they are down the stem.

observed phyllotactic patterns are optimal at harvesting light. Some other principle is at work.

This puzzle has proved irresistible to countless naturalists and scientists of all persuasions, many of whom have been convinced that the manifestation of such strikingly geometrical structures in nature reveals the operation of some deep-seated physical or mechanical process that imposes a kind of determinism on the biology of plant growth. It is fair to say that, while efforts to explain phyllotaxis have so far resisted any kind of scientific consensus, they have spawned something else entirely: a view of nature closely akin to that of the Platonists or the nineteenth-century German *Naturphilosophen*, pervaded by a liberal amount of geometric mysticism*. For it has been suggested that the mathematical patterns of phyllotaxis have far more profound significance than the mere fact of spiral ordering, and that at their root lie principles that govern our own sense of beauty and aesthetics. Phyllotaxis has become nothing less than the justification for a geometrization of nature.

But are we right to read so much into the shape and form of plants?

## CURVES OF LIFE?

Phyllotaxis is in many ways analogous to the body-patterning processes of animals that I have touched on already, and which I consider in more detail in the final chapter. It depends on breaking the cylindrical symmetry of the developing plant stem, just as the formation of limbs and organs in an embryo requires symmetry-breaking in a ball of cells. This spontaneous appearance of form baffled early scientists interested in phyllotaxis—as the English writer Theodore Cook explained in his 1914 book *The Curves of Life*: 'at the growing point of a plant where the new members are being formed, there is simply *nothing to see*'—a perfect articulation of the sheer surprise of symmetry breaking.

This 'growing point' at the tip of the stem is called the meristem. Here cells are multiplying rapidly, and just behind the advancing tip (the apex), side buds called primordia begin to protrude one by one. These will

---

*To Charles Darwin, attempting to explain these patterns 'could drive the sanest man mad.'

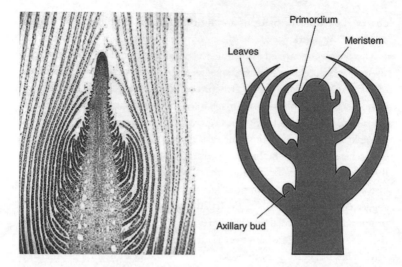

FIG. 6.2 The pattern of phyllotaxis is determined at the tip of the growing stem (called the meristem), where the leaf buds (primordial) are initiated. The photo on the left shows a meristem of the water weed *Elodea*. (Photo: Charles Good, Ohio State University at Lima.)

subsequently develop into leaves (Fig. 6.2). Successive primordia sprout at regular intervals, typically a day or so. The leaf pattern is determined according to where around the boundary of the apex the primordia appear. As the stem grows upwards, the positions of the primordia trace out a spiral when seen from above. One can see this arrangement more clearly by projecting the leaf positions on to a plane perpendicular to the stem, as shown for a monkey-puzzle branch in Fig. 6.3. Here the leaves are numbered from youngest to oldest, and lines are drawn through leaves that are in contact with one another. These trace out two spirals, twisting in opposite directions.

The double-spiral pattern is more immediately evident when the primordia develop not into leaves but into florets in a flower head, since in that case they may remain more or less all in the same plane. These florets are nothing more than modified leaves, and in the heads of sunflowers, daisies, and cauliflowers they generate captivating patterns (Fig. 6.4).

There appears to be something very strange and surprising about how the spirals are grouped. Travelling out along any one of the lines in Fig. 6.3, you

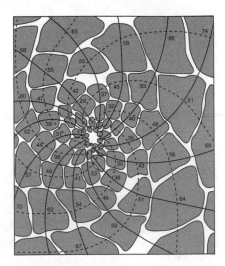

FIG. 6.3 The pattern of spiral phyllotaxis in the monkey puzzle tree. This is a projection of the leaf arrangement onto a flat plane, looking down the axis of the branch. Leaves are numbered consecutively from the youngest, and the two systems of spirals (solid and dashed lines) indicate leaves that touch one another.

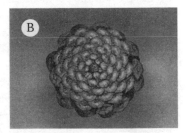

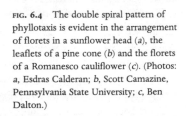

FIG. 6.4 The double spiral pattern of phyllotaxis is evident in the arrangement of florets in a sunflower head (a), the leaflets of a pine cone (b) and the florets of a Romanesco cauliflower (c). (Photos: a, Esdras Calderan; b, Scott Camazine, Pennsylvania State University; c, Ben Dalton.)

will find that the leaf numbers differ from one another by eight along the dashed lines and by thirteen along the solid lines. This construction allows a classification of the phyllotaxis pattern: it is denoted (8,13). Another way of expressing this relationship is simply to note that there are 8 spirals that turn clockwise, and 13 counter-clockwise.

Examples from other monkey puzzle branches show other phyllotactic relationships: (5,8), for instance, and (3,5). We find these same pairs if we count the spirals in a sunflower head, or in the leaflets of a pine cone. 'In many cones', says Thompson, 'such as those of the Norway spruce, we can trace five rows of scales winding steeply up the cone in one direction, and three rows winding less steeply the other way; in certain other species, such as the common larch, the normal number is eight rows in one direction and five in the other.' The Michigan botanist William Beal claimed in 1873 that 92 per cent of Norway spruce cones display the (3,5) arrangement.

To a mathematician, these pairs of numbers have a familiar ring. They are all adjacent integers in the sequence called the Fibonacci series, first defined in 1202 by the Italian mathematician Leonardo of Pisa, nicknamed Filius Bonacci or Fibonacci. Each term in the sequence is constructed by adding together the previous two, starting with 0 and 1. Thus, $0 + 1 = 1$, and the first three terms are 0,1,1. The next is $1 + 1 = 2$, then $1 + 2 = 3$, then $2 + 3 = 5$ and so on. The series runs 0,1,1,2,3,5,8,13,21,34 . . .

It is widely claimed that the phyllotaxis classifications of leaves, petals, or floret patterns in any plant species always correspond to pairs in this series. A corollary of this is that the number of petals on most flowers should be a Fibonacci number, since petals, like florets, are modified leaves. And indeed buttercups have five petals, marigolds have 13, asters 21.

Now this seems truly odd. Why should the form of plants be dictated by an abstract mathematical series? To the ancient Greeks it would have appeared entirely natural that number relationships should feature in the natural world. Platonists held a geometric conception of the universe, while the followers of Pythagoras considered numbers to be in some sense the building blocks of all things. Both philosophical schools considered that there exists a natural harmony of shape and form determined by simple ratios of proportion. That was certainly evident in music, where dividing a plucked string into lengths of ratios such as 1:2 and 1:3 made it resonate with notes that sounded in harmony with one another. We do not know if the Greeks recognized the Fibonacci series explicitly, but they knew of a most remarkable number that can be derived from it.

The ratio of successive terms in the Fibonacci series gets ever closer to a constant value the further one progresses along the series: $13/8 = 1.625$, $21/13 = 1.615$, $34/21 = 1.619$. This ratio approaches a value of $1.618034$ to the first six decimal places; to the Greeks it was known as the Golden Section. It has a value equal to $2/(\sqrt{5}-1)$, where $\sqrt{5}$ is the square root of 5, and it is commonly denoted by the symbol $\phi$ (phi). It seems a rather arbitrary number when expressed in decimals, but there is a simple and elegant way to interpret it geometrically. Suppose that you wish to divide a straight line into two unequal parts so that the ratio of the long part to the short is the same as the ratio of the whole line to the long part (Fig. 6.5a). The ratio of the two sections that satisfies this criterion is then $\phi$. Or if you wish to draw a rectangle that can be subdivided into a square and a smaller rectangle with the same proportions as the original one (but reduced in scale), the ratio of the two sides must again be equal to $\phi$ (Fig. 6.5b). These proportions were considered by the Greeks to be pleasing to the eye, and they are said to have based the dimensions of many temples, vases, and other artefacts on this ratio (it allegedly governs the proportions of the elevation of the Parthenon, although astute historians are sceptical). The Golden Section is also related to the logarithmic spiral (Chapter 1), which passes through the extremities of a series of rectangles growing in the successive proportions of the Fibonacci sequence (Fig. 6.5c). It is commonly held to be one of nature's 'special' numbers, like $\pi$ and $e$.

The Golden Section displays all manner of numerical quirks, and is undoubtedly a fascinating number;* but I can hardly emphasize firmly enough that in assessing its significance for 'nature's geometry' and aesthetics, we stand on the threshold of a numerological abyss. There is no end to the claims made for this proportion in nature and art. There is a long-standing idea, for example, that a perfectly proportioned human body has the ratio of the height of the navel to the total height equal to $\phi$. Likewise, the ratio of the length to width of the head, seen face on, is supposed to take this value, as are various proportions among the facial features. Now, it goes without saying that human bodies come in all shapes and sizes, and there

*For example, $1/\phi = \phi - 1$, $\phi 2 = \phi + 1$.

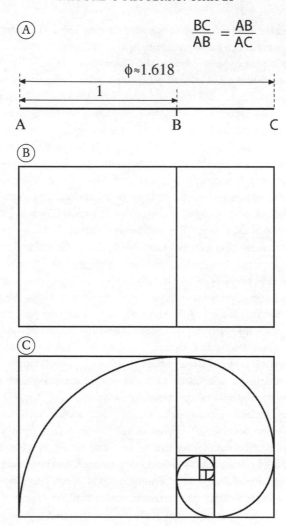

FIG. 6.5 The Golden Section dividing a line (a) and a rectangle (b). Quadrants drawn within successive squares of a recursively divided golden rectangle approximate a logarithmic spiral (c).

seems not the slightest reason to suppose that those with 'golden' proportions are any more attractive or in other ways privileged. The enthusiasm with which the abundant band of numerologists past and present, who I like to think of as the Golden Sect, seize on any evidence to support their contentions in this regard is perhaps best illustrated by the suggestion that we should be impressed by the vital statistics of film star Veronica Lake, who at one point in her career allegedly had a waist measurement of 21 inches and a chest of 34 inches.

What is unquestionable is that both the Golden Section and the Fibonacci series do crop up in art, but probably not on the scale that Theodore Cook, one of the most forceful proponents of the idea, would have us believe. (It was Cook who labelled the golden section 'phi', after the legendary Greek sculptor Phidias.) One can find all manner of grids drawn on classical ruins, medieval cathedral plans, and Renaissance paintings in support of this idea, usually with thick lines that preclude much numerical precision and which connect structural elements that do not obviously seem any more prominent than others that are ignored. It is precisely for this reason that the claims made for the Parthenon are impossible to substantiate. According to art historian Martin Kemp, 'there is no evidence that Renaissance and Baroque artists used such surface geometry in constructing their paintings'. If there is a 'da Vinci code', this is not it. The Golden Section did apparently become popular with later artists, such as Juan Gris and Georges Seurat, as well as with composers such as Bela Bartók, who seems explicitly to have built Fibonacci sequences into his scores. (The evidence that Johann Sebastian Bach encoded Fibonacci numbers into the *Art of Fugue* is more equivocal.) But all too often these forays into numerology by artists are taken as further evidence that $\phi$ has some universal, mystical significance, rather than as a sign of the self-fulfilling nature of that belief.

## SEARCHING FOR FIBONACCI

Most scientists agree with D'Arcy Thompson that one must seek an explanation of phyllotaxis not by recourse to the black box of adaptation, nor by probing the complexities of plant genetics, but by means of some *mechanical* process. On the whole, they have tended to accept the challenge at face value, and the various explanations that have been offered (some of which we will encounter shortly) are as ingenious and provocative as they are diverse.

But recently, cell biologist Todd Cooke of the University of Maryland took a thoughtful step back and asked the naive question that most people had considered to be already answered: just how good is the evidence for Fibonacci numbers in phyllotaxis anyway? When we are dealing with small numbers, it is all too easy to attribute significance to chance coincidences—that is precisely how numerology works. As Cooke says,

> The numbers 2, 3, and 5 (and their multiples) are frequently alleged to disclose the involvement of the Fibonacci sequence in a given process because they are taken to represent unique Fibonacci numbers as opposed to other 'non-Fibonacci' numbers. It follows from this allegation that any structure appearing in a group of five, such as the digits on the human hand or the petals of a rose flower, can be interpreted as being a manifestation of the Fibonacci sequence. But the first six positive integers are either components or multiples of the primary Fibonacci sequence; thus, a small group must be composed of at least 7 units before it appears to be unrelated to the primary Fibonacci sequence.

Even if we consider numbers up to 21, only four (7, 11, 17, 19) are not Fibonacci numbers or multiples of them. No wonder we see Fibonacci numbers everywhere! In fact, *all* positive numbers can in a sense be considered part of a Fibonacci series. We could, for instance, construct such a series by starting not with 1 and 1, but with 1 and 4:

1, 4, 5, 9, 14, 23 . . .

That may seem trivial—until you recognize that, as D'Arcy Thompson pointed out, the successive ratios of *all* such series converge to the Golden Section $\phi$. Try it if you don't believe it. As Thompson said, 'we must not suppose the Fibonacci numbers to have any *exclusive* relation to the Golden Mean.' Rather, this number emerges from what we might more usefully consider to be the Fibonacci *process* of creating a numerical series by summing the previous two members of the series. To Thompson, to linger too long over the mysteries of Fibonacci numbers in biology was to stand at risk of becoming a mystical Pythagorean.

Are we, then, justified in invoking the Fibonacci series and the Golden Section in any discussion of phyllotaxis? Let's first ask how persuasive a case there is for Fibonacci numbers, for example in the numbers of leaf-like elements (including petals) that appear around the stem of a plant. Whorls of two, three, and five petals are common in all flowering land plants (angiosperms)—both roses and flowers of the Ranunculaceae family (such as buttercups) typically have five, while lilies tend to have three or six. Although these are Fibonacci numbers (or multiples thereof), in view of what I've just said we have no good reason to suppose that this in itself implies any link with a 'Fibonacci' growth process. For flowers with larger numbers of petals, the evidence seems even more flimsy. Those of the Asteraceae family, which constitutes perhaps the largest group of flowering plants and includes chrysanthemums, daisies, artichokes, and sunflowers, are often said to have petal or floret numbers of 8, 13, 21, 34, 55, and 89. But Cooke simply looked in his back garden at the daisy-like perennial *Rudbeckia fulgida* and found that the flowers had between 10 and 15 petal-like florets, with an average of 12.8. That's close to the Fibonacci 13, but doesn't make much of a case for this number being particularly special. And for a variety of chrysanthemums bought from a local garden shop, Cooke found a range of between 20 and 36 florets, with an average (25.7) that did not seem particularly close to any number in the primary Fibonacci series.

Perhaps a more telling test of whether plants are genuinely governed by a Fibonacci process is to look at how the number of leaf-like features varies along a single stem. As we have seen, plants that display whorled phyllotaxis have rings of leaves that recur at regular intervals up the stem, with varying numbers of leaves per whorl depending on how broad the stem is when they start to form. Mare's tail (*Hippuris vulgaris*) is a classic example (Fig. 6.6). If these whorls were being formed by a Fibonacci process, we would expect to see leaf numbers in each whorl changing in jumps corresponding to the Fibonacci numbers, or at least their multiples. But a study on one variety of *Hippuris* showed that the leaf number increases in simple steps of 1: we find whorls with 5, 6, 7, 8, 9 leaves and so on. Moreover, these numbers seems to be simply related to the diameter of the meristem when the formation of the whorl begins. The wider the stem, the more room for leaves, with no special favour shown towards Fibonacci numbers.

FIG. 6.6   Whorls in mare's tail (*Hippuris vulgaris*). (Photo: Malcolm Storey.)

But what about spiral phyllotaxis—does this show Fibonacci numbers, as has been claimed? Here we seem on firmer ground. For example, a botanist named T. Fujita surveyed many angiosperms in 1938 and found that the great majority of those that display spiral arrangements of leaves do so according to a (2,3) pattern. Of the rest, most had a (1,2) or (3,5) pattern, four had a (5,8) pattern and just one an (8,13) arrangement.* In flower heads (botanists call these reproductive shoots, which contain the plant's reproductive organs, in contrast to the vegetative shoots of the

---

*In this system, the distichous structure shown in Fig. 6.1b corresponds to a (1,1) arrangement. Fujita apparently did not look at these at all.

leafy stems) the case is perhaps even more persuasive. Fibonacci spiral arrangements ranging from (1,2) to (34,55) are all found, such as the (13,21) patterns of stamens and carpels (male and female reproductive organs) in Japanese bigleaf magnolias (*Magnolia obovata*). (34,55) phyllotaxis is a rare special case, found in the florets of the sunflower *Heliathus annuus*; in fact, some particularly large sunflowers have been found to display patterns of up to (144,233). The Fibonacci spirals (8,13) and (13,21) are also seen in the arrangement of segments on the surfaces of pineapples.

What is more, when these spiral patterns undergo changes, for example by the budding of leaf primordia in plant stems at a higher rate in more mature plants, they pass Cooke's Fibonacci test by commonly making a leap from one Fibonacci pairing to the next in sequence. In other words, while we would be wrong to see any special significance in Fibonacci numbers of petals or leaves in a whorl, there does seem to be good reason to think that the growth process that leads to spiral phyllotaxis of leaves and leaf-like organs is governed by a patterning mechanism that favours the primary Fibonacci sequence. How, then, might such an arithmetical mechanism arise?

## MAKING SPIRALS

The German botanist Wilhelm Hofmeister proposed in 1868 that each new primordium appears periodically on the apex boundary in a position corresponding to the largest gap left by the preceding primordia: the primordia are simply trying to pack efficiently, just like atoms in a crystal. In 1904, Arthur Church, an English botanist, took this idea further in a book called *On the Relation of Phyllotaxis to Mechanical Laws*. It must be said that his was not the clearest of arguments, drawing in part on vague comparisons with spiralling vortices in fluid flow and magnetism and proposing that the 'energies of life' resemble electrical energy. Church was apparently a better artist than scientist—his botanical illustrations are things of real beauty—but as for his phyllotactic theory, D'Arcy Thompson sniffed magisterially that 'the physical analogies are remote, and the deductions I am not able to follow'.

As ever, Thompson had no time for organic mysteries of that ilk, but pointed out that spirals are an entirely predictable consequence of orderly packing of elements accompanied by growth in one direction: 'When the bricklayer builds a factory chimney, he lays his bricks in a certain steady, orderly way, with no thought of the spiral patterns to which this orderly sequence inevitably leads.' Nonetheless, Hofmeister and Church did help to establish the idea that phyllotaxis is related to the question of how new leaves can be packed on the meristem. This central notion has defined most modern thought about the phenomenon. The 'packing' thesis was apparently supported by computer calculations conducted by Helmut Vogel of the Technical University of Munich in 1979, which showed that the preferred angle of 137.5° allows optimal packing of primordia placed sequentially along a spiral, without ending up with lots of wasted space.

There is something special about this angle. Suppose we make the same 'golden' construction for a circle as I did earlier for a line, dividing it into a segment whose perimeter stands in the same ratio to the rest of the circle as the latter's perimeter does to the circumference of the original circle (Fig. 6.7). The angle subtended by the shorter segment—the angle at its apex—is called the golden angle. And guess what? It is equal to 137.5°.

This correspondence between the most common phyllotactic divergence angle and the golden angle was first identified by the mathematician brothers Louis and Auguste Bravais in 1837. However, it actually adds

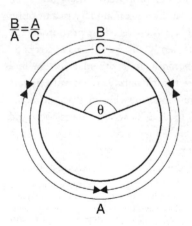

FIG. 6.7 Dividing up the circumference of circle into the Golden Section creates an arc that sweeps out an angle of about 137.5°— the 'golden angle'.

nothing to the Fibonacci revelations I have mentioned already: to say that leaves spiral up a stem with offsets of the golden angle is already to imply a relationship of phyllotaxis to the Fibonacci series. If you pick out points on a tightly wound spiral separated by an angle of 137.5° as measured from the centre, the human eye will always see two sets of counterposed spirals in the resulting array of points, the numbers in each set corresponding to successive Fibonacci numbers (how far into the series these pairs are depends on how tightly wound the generative spiral is). This fact was pointed out in 1907 by the Dutch botanist Gerrit van Iterson, and is lucidly explained by the mathematician Ian Stewart in his book *Nature's Numbers*. 'It all boils down to explaining why successive primordia are separated by the golden angle', Stewart says. 'Then everything else follows.'

But even if that angle can be explained by packing effects, we have not quite answered the mystery. In 1982 the mathematician J. N. Ridley of the University of the Witwatersrand in South Africa adapted Vogel's computer model to generate phyllotactic spirals with divergence angles differing slightly from 137.5°. What he found was that if this angle differed by no more than a tenth of a degree, the resulting packing of elements became rather loose and wasteful as the pattern grew larger (Fig. 6.8). But as Todd Cooke points out, it is untenable to imagine a biological system being able to organize itself with this degree of accuracy. And indeed, plants don't do that: the divergence angles of leaves in spiral phyllotaxis differ considerably both from one plant to another and from leaf to leaf on the same

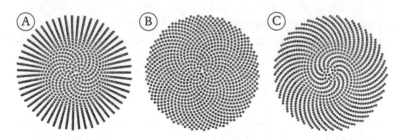

FIG. 6.8 Packing of objects in phyllotactic spirals that have divergence angles even slightly different from the golden angle of about 137.5° rapidly evolves into rather loose arrangements. Here the divergence angles are 137.45° (*a*), the golden angle (*b*), and 137.92° (*c*). (Images: Todd Cooke, University of Maryland, from Cooke, 2006.)

plant. One such analysis for the common cocklebur (*Xanthium pensylvanicum*) showed that the angles range from 124° to around 150°.

Packing effects, then, cannot fully explain Fibonacci patterns in phyllotaxis. Mathematically, tiny deviations from the golden angle demolish our pairs of Fibonacci spirals, whereas in real plants we can still see them clearly enough. So Cooke suggests that the mechanism of phyllotaxis must be one governed by other criteria, but which happens to give leaf arrangements that approximate those of truly optimal packing.

Several ideas have been proposed for what this mechanism might be. One of the most remarkable was put forward in 1992 by the French physicists Stéphane Douady and Yves Couder. They performed an experiment that seemed to have nothing in the slightest to do with botany: they dropped tiny droplets of a magnetic fluid onto a disk covered with a film of oil. The droplets floated on the oil layer. The apparatus sat in a magnetic field that polarized the magnetic droplets and caused them to repel one another. The researchers also applied a horizontal magnetic field, which was stronger at the periphery of the disk than at its centre—this pulled the droplets outwards towards the edge. Thus, as the droplets fell one by one, they were pushed out to the edges of the disk while repelling one another. This, the researchers argued, was a crude analogue of the formation of new primordia at the apex of a meristem, in which each is displaced off-centre by the next one to form.

It is unlikely that many botanists would have been persuaded that this was a particularly good mimic of the growth of a plant; but what happened was extraordinary. When the droplets were added at a fast enough rate, they travelled outwards to form Fibonacci double-spiral patterns ranging from (1,2) to (13,21) (Fig. 6.9), with successive droplets diverging at angles of close to 137.5°. And if the rate of droplet addition was lower, successive droplets diverged at 180° instead, giving a pattern that corresponds to distichous phyllotaxis. On the basis of this analogy, the exact kind of spiral pattern in phyllotaxis would be determined simply by the interval between budding of successive primordia, relative to the rate of stem extension. All very well—except that growing plants really are not magnetic droplets! Why would primordia repel one another?

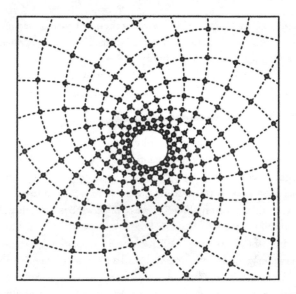

FIG. **6.9** Magnetic droplets moving from the centre to the edge of a circular dish while repelling one another trace out spirals of the same kind as those found in phyllotaxis. Yet here there are clearly only physical forces at play. (After Douady and Couder, 1992.)

It's time to bring back Alan Turing. We saw earlier that the pattern features of chemical Turing structures tend to steer clear of one another—in effect, they repel—because of the long-range inhibition that constitutes a crucial part of Turing's patterning mechanism. Could the same sort of inhibitory effect operate between budding primordia in phyllotaxis?

Turing himself suspected that might be so. He had a long-standing suspicion that his ideas on morphogenesis, which we encountered in Chapter 4, could shed some light on phyllotaxis too. This, after all, is simply a peculiarly mathematical type of morphogenesis. As such it fascinated him at least from his wartime years as a code-breaker at Bletchley Park, where he is recorded as studying daisies and fir-cones while off duty. And in a letter of 1951 to the English zoologist John Zachary Young about the theory of morphogenesis that he unveiled in his classic paper the following year, he wrote 'it will so far as I can see, give satisfactory explanations of ... leaf arrangement, in particular the way the Fibonacci series comes to be involved'. But in the paper itself he left that issue tantalizingly vague. His theory, he said, could explain how the circular symmetry of a stem or a cylindrical body such as a sea anemone might be spontaneously broken to form whorls of leaves, petals or tentacles, but it did not favour any particular number of pattern features. And yet in nature, as Turing wrote, 'the number five is extremely common, and the number seven rather rare'. He added that 'Such facts are, in the author's opinion, capable of explanation on the basis of morphogen theory ... [but] they cannot be considered in detail here.' That, he promised, was to come in a later paper—but this, entitled 'Morphogen theory of phyllotaxis', was still no more than a draft when he committed suicide.*

This is all the more tragic because even at that stage there existed enticing evidence of an inhibitory aspect to phyllotaxis that Turing could have drawn on. In the 1930s the botanists Mary and Robin Snow at Oxford University showed that new leaf primordia require a certain minimum of space on the side of the conical meristem before they can begin to form. In other words, only when the apex of the stem has grown far enough ahead can a fresh bud appear on the slope below: the apex *inhibits* primordium formation within a certain distance. This suggests that there might be some diffusing morphogen(s) involved in primordium formation at the stem tip that creates an inhibitory effect.

In 1977 the British botanist J. H. M. Thornley postulated that phyllotaxis might be governed by what we can regard, within Turing's scheme, as an *activating* morphogen. He suggested that such a morphogen must accumulate at the tip to a certain threshold level before primordium formation can commence. In fact, the Snows identified such a candidate substance in the 1930s, when they found that applying a paste containing the plant

---

*This draft has now been published in *Morphogenesis: Collected Works of A. M. Turing*, vol. 3, ed. P. T. Saunders (North-Holland, 1992), and it is discussed by Jonathan Swinton in *Alan Turing: Life and Legacy of a Great Thinker*, ed. C. A. Teuscher (Springer, 2004). Turing's basic idea is akin to that discussed below, in which a reaction–diffusion system acting at the growing tip of the plant lays down spots that subsequently grow into buds on the cylindrical stem. Turing expanded on this theory in another draft manuscript, 'Outline of the development of a daisy', which has been published in the same collected edition of his works.

hormone auxin to the meristem of lupins initiated the growth of a new leaf bud. This manipulation of phyllotaxis could even alter the qualitative pattern, changing it from decussate (Fig. 6.1c) to spiral. Later it was found that other chemical compounds that interfere with auxin, for example by slowing its diffusion rate through the plant tissue, can also transform the phyllotactic pattern. Perhaps auxin is an activator of primordium growth, and its depletion by some sink produces inhibition?*

This idea was confirmed in 2003 when a team of European biologists showed that phyllotaxis is regulated by proteins that ferry auxin through the outer 'skin' of the meristem up towards its apex. They found that existing leaf primordia soak up auxin and thus act as sinks, inhibiting the formation of any new buds nearby. But that alone, the researchers concluded, would seem to favour a distichous pattern in which new primordia always appeared on the opposite side of the stem, as far away from the previous primordium as possible. How, then, do the spiral divergence angles arise?

It seems that was essentially what Turing was aiming to explain. Now Hans Meinhardt and his colleague André Koch have done the job instead. They have proposed a reaction–diffusion model in which certain hormones act as inhibitors that repress primordia formation in a given region of the stem until the tip has grown far enough for the hormone concentration there to fall below a threshold value. Once this long-ranged inhibition becomes sufficiently weak, some local activator molecules switch on cell proliferation to induce the budding of a primordium.

In their model, a *second* activator–inhibitor mechanism controls the angle between successive primordia. Meinhardt and Koch calculated the primordia patterns the model generates on an idealized plant stem modelled as a narrow, hollow cylinder. They found that the primordia became positioned along the stem in a (2,3) spiral phyllotaxis pattern (Fig. 6.10). By making some simple and reasonable assumptions about how cells differentiate around the primordia, Meinhardt and Koch were even able to account for the formation of the little 'secondary' structures called axillary

---

*Auxin boosts plant growth, and both it and various related synthetic chemical compounds are used in agriculture for that purpose. But an overdose of auxin can kill plants instead, and so these compounds are also used and herbicides—including the notorious Agent Orange.

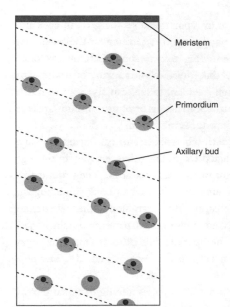

Meristem

Primordium

Axillary bud

FIG. 6.10   Spiral phyllotaxis can be generated in a reaction–diffusion model of patterning on a cylindrical plant stem, here shown rolled out into a flat sheet. The spiral sequence of primordia is indicated by dashed lines. New primordia develop below the meristem at the top of the cylinder. (After Koch and Meinhardt, 1994.)

buds seen just above the developing leaf where it joins the stem in real plants. Whether the model can generate higher-order Fibonacci spirals is not clear, however, and nor is there yet any evidence that such a double activator–inhibitor process really operates in plants. The model hints that reaction–diffusion could indeed do the job—but that is all.

## UNDER PRESSURE

A Turing mechanism for phyllotaxis thus looks quite plausible in terms of what we know about plant biochemistry. But it's not the only game in town. Some researchers think that phyllotactic patterns might be produced *mechanically* in the stem tip. Plant biologist Jacques Dumais and mechanical engineer Charles Steele at Stanford University in California suggested in 2000 that the initiation of new primordia could result from the buckling of the skin, or tunica, of the meristem surface due to the compression it experiences as it grows. The tunica is soft and squishy at the apex but gradually hardens towards the edges, making it prone to buckling as new growth deforms it. In support of this idea, Patrick Shipman and Alan Newell of the University of Arizona in Tucson have calculated what the buckling would look like in simple shells with this structure. They found that in the lowest-energy patterns—those that formed most readily— intersecting wrinkles divide the tunica into a series of little peaks positioned in Fibonacci spirals (Fig. 6.11a, c). Under some circumstances the buckling could instead produce a distichous pattern (Fig. 6.11b, d).

Physicists Zexian Cao and colleagues at the Chinese Academy of Sciences have supplied a remarkable demonstration of such stress-driven Fibonacci patterning. They made hot microscopic blobs of silver coated with silicon oxide, and watched what happened to the hard shell as the blobs cooled down. Because the silver core contracts fastest, the shell becomes

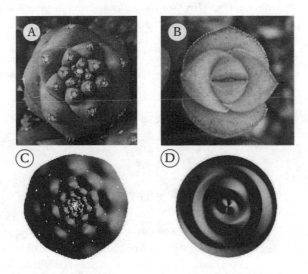

FIG. 6.11   Phyllotactic patterns can be reproduced in a model of buckling of the elastic skin or tunica of the meristem. Shown here are spiral (a) and decussate (b) patterns in cacti, and the corresponding structures produced by the buckling model (c, d). (Photos and images: Patrick Shipman, University of Maryland, and Alan Newell, University of Arizona, from Shipman and Newell, 2004.)

compressed and laced with lines of stress. This web of stress lines was made visible by the condensation of new, smaller nodules of material on the shell surface, since these smaller structures grew more readily at high-energy, stressed locations. On nearly spherical blobs, the nodules became organized into regular hexagonal arrays. But if the blobs had a slightly conical shape, rather like that of a plant meristem, the nodules tended to form Fibonacci spirals such as (13, 21) (Fig. 6.12).

It is not yet clear whether a Turing or a stress mechanism best accounts for phyllotactic patterns. What seems more certain is that stresses in growing plant tissues govern another aspect of their patterned shapes: the frond-like crinkles of some leaves and flowers. These are found, for example, in the petals of orchids, the leaves of lettuces and ornamental cabbages, and the fronds of seaweed (Fig. 6.13a,b). In daffodils such wrinkles complicate the edge of the flower's central trumpet (Fig. 6.13c). Where do the ripples come from?

They can be mimicked by simply tearing apart a piece of plastic sheeting, such as that used for lining waste bins (Fig. 6.14). It is striking here that, although the mechanical process used to generate the patterns is very simple, the shape is very complex. In fact, in some cases it can be seen to have ripples on ripples on ripples at increasing

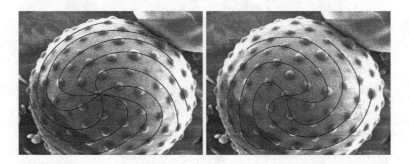

FIG. 6.12 Fibonacci spiral patterns equivalent to those seen in phyllotaxis are generated by spherules grown on a shell of silicon oxide coating a conical blob of silver with a rounded tip, shaped like a plant's meristem. The patterns here are caused by stresses induced in the surface as the hot tip cools, which create preferential sites for spherule formation. The tip shown here has a (13,21) spiral pattern. (Photos: Cao Zexian, Chinese Academy of Sciences, Beijing. From Li *et al.*, 2005.)

FIG. 6.13  The edges of many plant structures, such as orchid petals (a), ornamental cabbage leaves (b), and daffodil heads (c), are decorated with ripple-like fronds. (Photos: a, Stef Yau; b, gemteckı; c, Peter Allen.)

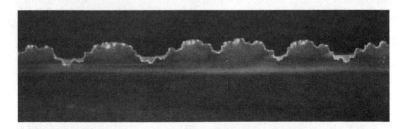

FIG. 6.14  Wrinkles in torn plastic have a characteristic wavy structure at several size scales. (Photo: Eran Sharon, kindly supplied by Michael Marder, University of Texas at Austin.)

magnifications, giving the pattern the characteristic property of forms known as fractals, which we shall encounter in Book III.

Making ripples in torn plastic sheeting helps us to see where they come from. In tearing the sheet, we first stretch it, deforming the plastic permanently. The stretching happens only around the edges of the tear—elsewhere the sheet is unstretched. This means that the stretched portions cannot simply stay flat while increasing their length—the 'extra' length has to be accommodated in some other way, by buckling out of the flat plane of the sheet. So the crinkles are the sheet's response to having become 'too long' in this region to fit smoothly into the end-to-end space it is constrained to occupy. What this means, however, is that the sheet's shape no longer reflects the symmetry of the forces that generated it. Whereas we just pulled at the plastic in the plane of the sheet, it has adopted a shape that is non-symmetrical *out of* the plane. In this sense, the process again involves symmetry-breaking.

Michael Marder of the University of Texas at Austin and his co-workers have shown that the crinkled edges of leaves and flowers can likewise be understood by assuming that the plant tissue here has become 'too long' to be contained within its boundaries. That is not caused by stretching as such, but simply by uneven growth rates of the tissues in different parts of the initially flat structures. Marder's team was able to induce ripples at the edge of normally flat aubergine (eggplant) leaves, by applying auxin along these edges. As we have seen, auxin is a hormone that controls plant growth—in this case, it simply speeded up the rate of leaf tissue growth. After several days, the treated leaves developed wavy edges (Fig. 6.15). Marder says that, given that leaves have curved and often rather complex shapes, it is actually more surprising that many leaves are *not* crinkled at the edges—that they somehow manage to adjust their tissue growth rates to ensure flatness. It appears that this is done through genetic regulation: genes act to fine-tune the growth rates so that parts of the leaves don't outpace others. It is no wonder, then, that a telltale sign of some plant diseases is crinkled leaves, which signal disruption of the normal genetic controls on growth.

Marder and his co-workers calculated what the buckling patterns will be for strips of thin material with a gradually varying total length from one

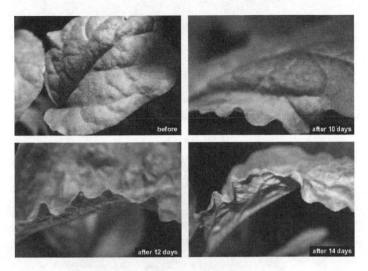

FIG. 6.15 Wrinkles are induced at the edge of a normally flat leaf (an aubergine plant) by administering the hormone auxin, which alters the growth rate of the tissue. (Photo: Michael Marder, University of Texas at Austin.)

edge to the other.* They found that indeed the ripples can display the hierarchical 'waves upon waves' patterns seen in plants (Fig. 6.16a). Applied to the edge of a cylinder, these deformations produce the delicate and rather beautiful patterns familiar in daffodils (Fig. 6.16b).

Buckling might conceivably also explain the surface patterning of some fruits and vegetables, such as pumpkins, gourds, melons and tomatoes. These have soft, pulpy flesh confined by a tougher, stiffer skin. Some fruits have smooth surfaces that simply inflate like balloons as they grow, but others are marked by ribs, ridges or bulges that divide them into segments (Fig. 6.17a). According to Xi Chen of Columbia University in New York, working in collaboration with Zexian Cao in Beijing and others, these shapes could be the result of buckling.

---

*Technically, this variation is caused by changing the so-called *metric* of the space that the strip occupies—it is rather like distorting the space itself, as though it was a rubber sheet. Such changes in the metric of space were used by Einstein in his theory of general relativity to understand the effects of gravity. On page 260 I discuss D'Arcy Thompson's use of this elastic distortion of space for making anatomical comparisons across species.

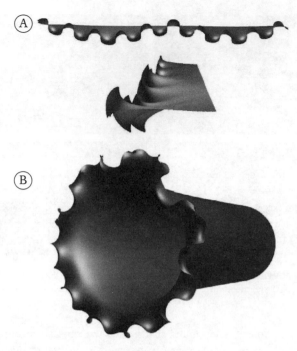

**FIG. 6.16** The ripple patterns calculated in a model of buckling at the edge of an elastic sheet, for a strip (a) and a cone (b). The latter resembles the head of a daffodil. (Images: Michael Marder.)

**FIG. 6.17** Many fruits have ribbed bodies (a). This might be explained by the mechanical buckling of their outer skin, which creates distinctive patterns on spheroidal shells (b). (Photos and images: Xi Chen, Columbia University.)

This is a familiar process in laminates that consist of a skin and core with different stiffness: think, for example, of the wrinkling of a paint film stuck to wood that swells and shrinks. Under carefully controlled conditions, this process can generate patterns of striking regularity (Figure 6.18).

Chen and colleagues performed calculations to predict what will happen if the buckling occurs not on a flat surface but on spherical or ovoid ones (spheroids). They found well-defined, symmetrical patterns of creases in a thin, stiff skin covering the object's surface, which depend on three key factors: the ratio of the skin thickness to the width of the spheroid, the difference in stiffness of the core and skin, and the shape of the spheroid—whether, say, it is elongated (like a melon or cucumber) or flattened (like a pumpkin).

The calculations indicate that, for values of these quantities comparable to those that apply to fruits, the patterns are generally either ribbed—with grooves running from top to bottom—or reticulated (divided into regular arrays of dimples), or, in rare cases, banded around the circumference (Figure 6.19). Ribs that separate segmented bulges are particularly common in fruit, being seen in pumpkins, some melons, and varieties of tomato such as the striped cavern or beefsteak. The calculations show

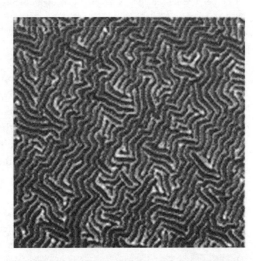

FIG. 6.18 The wrinkles that appear in a flexible film compressed by the shrinking of an underlying substrate can have startling regularity, as seen here in a thin metal film coating the surface of a slab of rubbery polymer. (Photo: George Whitesides, Harvard University, from Bowden et al., 1998.)

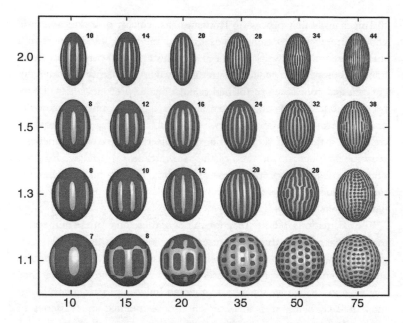

FIG. 6.19 The buckling shapes of spheroidal shells have a geometric regularity with patterns that depend on the lenght-to-width ratio (varying from top to bottom) and the ratio of the shell thickness to the object's diameter (varying from left to right, with thinner shells on the right). The numbers next to each shape indicate how many ridges or grooves they have. (Image: Xi Chen, Columbia University, from Yin et al., 2008.)

that spheroids shaped like such fruits may have precisely the same number of ribs as the fruits themselves (Figure 6.17b).

For example, the 10-rib pattern of Korean melons remains the preferred state for a range of spheroids with shapes like those seen naturally. That's why the shape of a fruit may remain quite stable during its growth (as its precise spheroidal profile changes), whereas differences of, say, skin thickness would generate different features in different fruits with comparable spheroidal forms.

Chen suggests that the same principles might explain the segmented shapes of seed pods, the undulations in nuts such as almonds, wrinkles in butterfly eggs, and even the wrinkle patterns in the skin and trunk of elephants. So far, the idea remains preliminary, however. For one thing, the mechanical behaviour of fruit tissues hasn't been measured precisely enough to make close comparisons with the calculations. And the theory makes some unrealistic assumptions about the elasticity of fruit skin. So it's a suggestive argument, but far from proven. Besides, Chen and his colleagues admit that some of the shaping might be influenced by subtle biological factors such as different growth rates in different parts of the plant, or direction-dependent stiffness of the tissues. They argue, how-ever, that the crude mechanical buckling patterns could supply the basic shapes that plants then modify. As such, these patterns would owe nothing to evolutionary fine-tuning, but would be as inevitable as the ripples on a desert floor.

I daresay Figure 6.18 may have already put you in mind of another familiar pattern too. Don't those undulating ridges and grooves bring to mind the traceries at the tips of your fingers (Fig. 6.20)? Yes indeed; and Alan Newell has proposed that these too might be the product of buckling as the soft tissue is compressed during our early stages of growth. About ten weeks into the development of a human foetus, a layer of skin called the basal layer starts to grow more quickly than the two layers (the outer epidermis and the inner dermis) between which it is sandwiched. This confinement gives it no option but to buckle and form ridges. Newell and his colleague Michael Kücken have calculated what stress patterns will result in a surface shaped like a fingertip, and how the basal layer may wrinkle up to offer maximum relief from this stress.

The buckling is triggered and guided by tiny bumps called volar pads that start to grow on the foetal fingertips after seven weeks or so. The shape and positions of volar pads seem to be determined in large part by genetics—they are similar in identical twins, for instance. But the buckling that they produce contains an element of chance, since it depends on (among other things) slight non-uniformities in the basal layer. The American anatomist Harold Cummins, who studied volar pads in the early twentieth century, commen-ted presciently on how they influence the wrinkling patterns in ways that cannot be fully foreseen, and which echo universal patterns elsewhere: 'The skin possesses the capacity to form ridges, but the alignments of these ridges are as responsive to stresses in growth as are the alignments of sand to sweeping by wind or wave.' Newell and Kücken found that the shape of the volar pads govern the print patterns: if they are highly rounded, the buckling _

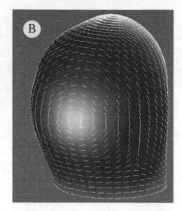

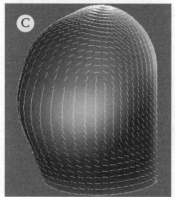

FIG. 6.20 Fingerprint patterns (a) seem to be generated by buckling of skin at the fingertips. The buckling pattern—concentric or arched— is apparently controlled by the shape and position of tiny bumps called volar pads, as indicated in a mathematical model that produces whorls (b) or arches (c). (Images: Michael Kücken, Technical University of Dresden, and Alan Newell, University of Arizona; from Kücken and Newell, 2004.)

generates concentric whorls, whereas if the pads are flatter, arch-shaped ridges are formed (Fig. 6.20b, c). Both of these are seen in real fingerprints.

Again, then, here are patterns that do not depend on any genetic encoding of the sort that says 'put a ripple here', but are instead the spontaneous response to a set of physical forces and constraints—the kind of process championed by D'Arcy Thompson. In the final chapter I shall consider the central and most challenging question behind Thompson's thesis: how far can this spontaneous patterning go in producing an entire living organism from a featureless embryo?

# 7

# UNFOLDING THE EMBRYO

### The Formation of Body Plans

Alan Turing's speculations about animal markings and phyllotaxis were a natural corollary of a theory that showed how uniform space might become spontaneously patterned into more or less regular structures. But they were incidental to his main objective, which was to explain the development, in its earliest stages, of forms far more complex. His sights were set on nothing less than the human body: an intriguing blend of regularity, symmetry, and apparently arbitrary form. Looked at in purely geometric terms, the body is a remarkable and puzzling thing. We possess little of the crystalline precision of a diatom shell, and yet we are anything but amorphous. There is the obvious bilateral symmetry; but the regularities go well beyond that. The bone structure of the arms echoes that of the legs, each ending in a series of five nearly-but-not-quite-identical digits. The backbone is segmented into nearly identical units, our ribs are a rack of struts, our hairs sit in equidistant follicles.

Does this mean the human body has *pattern*, in the sense that I have defined it? Let us not get too semantic about it; one can surely say that, at the least, the fine balance of the regular and the irregular contributed to the fascination that the human form held for scientists ranging from the comparative zoologists of the eighteenth century to a mathematician such as Turing. D'Arcy Thompson considered it an error of biology to believe that 'the peculiar aesthetic pleasure with which we regard a living thing is somehow bound up with the departure from mathematical regularity'—as though such a departure was a hallmark of life. Yet he felt compelled to admit that

> The organic forms which we can define more or less precisely in mathematical terms, and afterwards proceed to explain and to account for in terms of force... are few in number compared with Nature's all but infinite variety... For one reason or another there are very many organic forms which we cannot describe, still less define, in mathematical terms.

In other words, it was one thing to develop a theory of the Nautilus shell, the flower stem, or the tunicate; but could there ever be a mathematical theory of D'Arcy Thompson?

Thompson found an ingenious way to attack this problem of the apparent mathematical intractability of complex biological form. He demonstrated that comparable forms, such as the shapes of different fish or of primate skulls, might be transformed one into another by drawing them on grids which were then deformed like rubber. It was an attractive way to make zoological comparisons, but it was otherwise nothing more than a nice descriptive tool which, as Thompson himself admitted, served merely to render with a little more precision Aristotle's comment that related species differ from each other simply in terms of the proportions of their parts. Thompson's morphological mapping only highlights, rather than addresses, the key underlying question: why is it that species share these features in common? Where do these forms come from and how are they altered over evolutionary time so that, say, the fins of fish become arms and legs, which in turn become wings?

Until relatively recently, biologists felt that the answer was provided by the so-called Modern Synthesis of genetics and evolution that was drawn up in the 1930s and 1940s. Small, random changes in the nature of a species' genes led to small variations in physical attributes, on which selective pressure then acted to amplify advantageous mutations. The question of precisely how the genes generate those attributes in the first place was almost entirely open, but was considered a mere detail that geneticists and molecular biologists would thrash out in due course.

We now know that the issue of causative mechanism—the mode of action of the genetic machinery that makes an organism—is not a detail at all. Discoveries since the mid-1980s have revealed that the Modern Synthesis is simplistic if not downright wrong in some regards, and that the genetics of morphogenesis, of biological shape and form, are more remarkable than anyone had guessed.

It would be satisfying to be able to say that Turing's hypothesis provided the missing link between biomolecules and biological form. Some modern accounts of morphogenesis come close to giving that impression. But that is not the case; Turing's theory has undoubtedly found a place in developmental biology, as we have seen already, but it is not the key to the appearance of form in embryogenesis that he had hoped it might be. Yet some elements of Turing's theory have proved exceedingly prescient. And it is fair to say that the emerging understanding of morphogenesis is one that pays fresh heed to questions of patterning and symmetry breaking. Our bodies are not exactly spontaneously patterned from top to toe in the manner of a lump of agate, or even of a leopard's pelt; but neither are we constructed according to instructions that assign each cell its place and function from the outset. Rather, we are—in a sense that is almost alarmingly exact—each an example of the kind of toolkit pattern that we have encountered already in the butterfly's wing: a delicate combination of the preordained and the contingent.

## WHICH CAME FIRST?

Not all organisms, of course, grow into bigger versions of their youthful selves, variously elaborated or truncated. The case of the butterfly is celebrated in countless children's tales, but how perplexing this must seem to children: an adult that bears scarcely any resemblance to the infant.* In *Natural Theology*, William Paley pointed to this very puzzle. After suggesting that life exists to pass on organization in which the new is related to the old, he admits that

---

*Julia Donaldson's charming *Monkey Puzzle* plays on that very confusion, as a butterfly directs a monkey separated from its mother to all manner of inappropriate creatures, unaware that most mothers have the same general shape as their children.

There are other cases, especially in the progress of insect life, in which the dormant organization does not much resemble that which incloses it, and still less suits with the situation in which the inclosing body is placed, but suits with a different situation to which it is destined. In the larva of the libellula, which lives constantly, and has still long to live, under water, are descried the wings of a fly, which two years afterwards is to mount into the air.

For Paley this was simply an illustration of God's foresight and ability to 'mold and fashion the parts of material nature, so as to fulfill any purpose whatever which he is pleased to appoint'. That is a possible but unnecessary hypothesis, and has nothing to do with science. The advent of genetics sharpens this question, for it makes us demand how a butterfly, equipped with the same genes as the caterpillar from which it grew, acquires such an utterly different body plan.

In the eighteenth century no one was troubled by the question of how babies grow from embryos, because it was assumed that (with the occasional exception) creatures start life as miniature but fully formed versions of their adult selves, and just grow bigger. People, it was thought, grow from microscopic homunculi in the womb, which possess formative arms, legs, eyes, and fingers (even if we cannot see them). The problem with this idea is that it seems to entail an infinite regression: unless you are prepared to accept the formation of pattern from a shapeless egg at some stage, you have to assume that the female homunculi contain even smaller homunculi in their tiny ovaries, and so on for all future generations.

This notion of 'preformation' is easy to ridicule, but the American biologist Stephen Jay Gould has pointed out that, so long as we do not reduce it to a caricature, there are perfectly good reasons why it might have seemed plausible to eighteenth-century scientists. And at least it offered a mechanistic interpretation of form, whereas the alternative seemed to be a quasi-mystical appeal to some morphogenetic force of nature that wrought form out of shapeless living matter. In this view, the embryo is imbued with an invisible pattern that finds gradual expression as a mature organism.

That idea made a good fit with the *Naturphilosophie* of German Romanticism, the intellectual foundation of Ernst Haeckel's theory of morphology. Haeckel, you may recall, proposed that the embryonic development of a 'higher' organism recapitulates its evolutionary history, the embryo passing through forms in which lower organisms have become 'stuck'. The eighteenth-century French philosopher Jean-Baptiste René Robinet put it eloquently, and also revealingly, when he described such simple animals as 'the apprenticeship of nature in learning to make man'. Although Haeckel's phrase 'ontogeny recapitulates phylogeny' was the clearest and most influential expression of this idea, it was more or less an inevitable consequence of *Naturphilosophie*, which some have traced to Goethe; certainly, Haeckel was merely refining an old idea. It stems from the Romantic belief in the unity of nature, according to which man is linked not only to other animals but to plants and minerals. That image is very much back in vogue today, although we feel far less comfortable with the other assumption of the *Naturphilosophen*: that of a hierarchy of being in which humankind comes at the top, representing the deterministic goal of the creative potential of nature. Given that belief, it is not hard to see why Haeckel and his contemporaries felt justified in imposing such a hierarchy *within* the human race too. They were hardly alone in that, and we risk being ahistorical if we condemn them too harshly; but Haeckel, as we have seen, went further than most, and has been justly accused of 'evolutionary racism'.

One might think that Haeckel's celebrated 'biogenetic law' does not actually explain anything, merely asserting that one barely understood process resembles another. But Haeckel himself saw this as a causative mechanism. 'Phylogenesis', he wrote, 'is the mechanical cause of ontogenesis'. What can this mean? While there are grounds for regarding Haeckel as a historical determinist, he is not a crude vitalist—he understood that somewhere in the explanation for how organisms get their forms there must be a role for molecular physiology. And indeed he wrote in 1866 that

> Phylogenesis ... is a physiological process, which, like all other physiological functions of organisms, is determined with absolute necessity by mechanical causes. These causes are the motions of the atoms and molecules that comprise organic material ... Phylogenesis is therefore neither the foreordained purposeful result of an intelligent creator, nor the product of any sort of unknown mystical

force of nature, but rather the simple and necessary operation
of . . . physical-chemical processes.

One cannot imagine that D'Arcy Thompson would have found much to
quibble with in this eminently rational assertion. But it leads you to expect
that Haeckel has something in mind, some process by which those 'atoms
and molecules' arrange themselves into people. If he did, he never disclosed
what it was. Rather, he merely played this same tune again and again—as
Gould puts it, 'This theme is invoked in all his popular works with an ardour
and insistency that demands assent by sheer repetition, rather than by any
increment of profundity.' Darwin was among those who despaired of
Haeckel ever explaining his 'mechanism' of phylogenesis, generously own-
ing that 'Perhaps I have misunderstood him . . . His views make nothing
clearer to me, but this may be my fault.'

Of course, Darwin had no molecular mechanism that accounted for
evolution either; but he was more ready to acknowledge the deficit.
Haeckel seemed to feel that scientists should be satisfied with his vague
pronouncements on 'the motions of atoms and molecules', and he heaped
scorn on those who sought for genuine mechanistic origins of the shape
and form of embryos. One of his most severe critics, the German biologist
Wilhelm His, complained with good cause that the biogenetic law in itself
answered nothing: 'An array of forms, following one after the other is
really, and this must be emphasized again and again, no explanation.'
Whereas in former centuries, His grumbled, 'authors professed to read in
every natural detail some intention of the *creator mundi*, modern scientists
have the aspiration to pick out from every occasional observation a
fragment of the ancestral history of the world'.

His anticipates D'Arcy Thompson in seeking proximate, mechanical
causes for the shapes of embryos. He argued that differences in the rates of
tissue growth created mechanical pressures that lead to buckling and
folding of the elastic sheets, much as they do in the case of leaves, petals,
and fingertips as described in the previous chapter. He drew comparisons
between embryonic organs and the shapes of deformed rubber tubes.
Thus, while heredity might account for the differences in tissue growth,
the morphological results were a matter of mere mechanics. Haeckel
treated His's theory with contempt, not because he could identify any

real flaws in it, but because (he felt) it was obviously wrong since it ignored phylogeny—and because it smacked of something as ignoble as engineering. 'He imagines constructive Nature to be a sort of skilful tailor', Haeckel sneered. 'The ingenious operator succeeds in bringing into existence ... all the various forms of living things by cutting up in different ways the germinal layers, bending and folding, tugging and splitting.' It is, he concluded, 'a sartorial theory of embryology'.

There is a broader lesson in this dispute. For I imagine it is clear by now that the real argument is about approaches to doing science. There are those who seek for the big picture, for the relationships between things; and there are those who are not content until the details are spelled out. The former tend to be dismissive of reductionist questions about *how* precisely one thing leads to another—they lay out the whole canvas, and leave it to apprentices and clerks to fill in the details. The latter think like engineers, remaining unconvinced that the whole edifice will stand unless you have built it up brick by brick. Science needs both types, but it can be hindered by the shortcomings of each. Grand visions like those of Haeckel generate questions that act as stimuli to further research, and offer some kind of framework, however temporary or shaky, that prevents science from fragmenting into a welter of tiny facts. But it is only when you pay attention to the details that you can be sure you are on the right track. Darwin was, of course, a Grand Visionary, albeit one who was able to resist the dogmatic certainties that frequently bedevil such people. D'Arcy Thompson was the engineer, wary of handwaving generalizations and insistent on the need for mechanical causes.

In any event, Haeckel's biogenetic law is wrong: there is no reason to think, and indeed now plenty of reason to deny, that a growing human embryo passes through stages that bear any direct connection to the mature adult forms of human ancestors throughout evolutionary history. Many objections were raised to the theory in Haeckel's own time, both on phylogenetic and physiological grounds, but none of them was deemed fatal, and the theory enjoyed at least three decades of wide acceptance. Stephen Jay Gould argues, however, that the demise of Haeckel's law owed less to the emergence of the science of genetics (which raised genuine scientific problems with it) than to the vicissitudes of scientific fashion: embryologists became more interested in looking at the question of

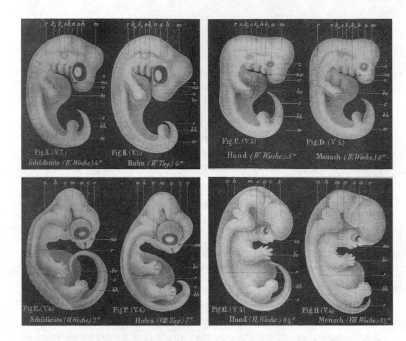

FIG. 7.1 Ernst Haeckel based his 'biogenetic law' (the idea that the development of an embryo recapitulates evolutionary history) on alleged similarities between the early stages of embryogenesis in diverse species, as illustrated here in his *Natural Hsitory of Creation* (1870). The veracity of his drawings was later questioned.

morphogenesis and development from an experimental point of view than in seeking recourse in analogy, comparison, and generalization.

Yet there is one particular aspect of the critique of Haeckel that is worth dwelling on. Haeckel, as we saw at the start of this book, attached great importance to the visual documentation of an argument. He did not simply use images as supporting evidence for his ideas, as scientists did then and still do now; rather, he considered that his arguments could be made in visual terms. To validate the biogenetic law, he showed the embryonic stages of various organisms alongside one another, demonstrating their initial resemblances (Fig. 7.1). These illustrations were prepared by Haeckel himself from his own microscopic observations and those of others. But his critics, including Wilhelm His, accused him of distorting these images to accentuate the similarities. To today's scientists this looks like the heretical sin of manipulating the data, and it is apt to put Haeckel in very bad light with those who agree that the drawings represent a bowdlerized version of what could actually be seen. Others suggest in his defence that the pictures were meant to be no more than schematic, and in that sense they are no different from any other scientific illustration that emphasizes what the viewer is meant to see. Moreover, scientists in those days did not have such a prohibition against a bit of 'tidying' of the visual data (His, too, has been accused of some sloppiness with his illustrations).

One could counter that defence with the charge that the naturalism of the drawing style gives scant warning to the reader not to make too literal an intepretation. But beyond even that, there do seem to be signs of more deliberate manipulation. For example, when Haeckel drew the embryonic development of an echidna based on the illustrations made by zoologist Richard Semon in 1894, he removed the limbs from the early stages so that he could claim 'There is still no trace of the limbs or "extremities" in this stage of development . . . [this] proves that the older vertebrates had no feet.'

The degree to which Haeckel's fudging of the data should be regarded as scientific fraud is still debated today. What is more clear, however, is that the way he chose to present the visual evidence for the biogenetic law should make us a little cautious about the representations of symmetry and order in the natural world depicted in *Art Forms of Nature*.

## STRIPED EGGS

The development of an embryo can now be followed in unambiguous detail with modern microscopes. It is a systematic and predictable process in which an initially featureless ball of cells is progressively segmented by grooves and other structures until the rudiments of the body plan start to appear. Embryogenesis has been particularly well studied in the fruit fly *Drosophila melanogaster*, an organism complex enough to be considered directly relevant to human development but simple enough, and with a short enough life cycle, to investigate in the laboratory. During the first half day of embryonic growth, the ovoid egg can be seen to become

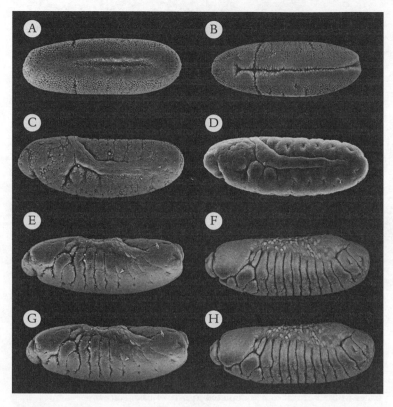

FIG. 7.2 A fruit fly embryo develops grooves that become compartments of its segmented body. The sequence shown here covers just 12 hours. (Photos: Rudy Turner, Indiana University, kindly provided by Sean Carroll, University of Wisconsin-Madison. From Carroll, 2005.)

grooved along and perpendicular to its long axis, creating structures that will later become the body segments (Fig. 7.2). In other words, the form of the adult fly does not simply emerge in rough outline from the smooth egg; instead, its symmetry is broken gradually. Morphogenesis is a matter of gradual elaboration, and in the early stages seems to offer few clues about the eventual destination.

But microscope images like this tell only part of the story. There are patterns forming even before we see their expression in terms of tissue folds and grooves, and these can be revealed by adding dyes or fluorescent marker molecules to the embryo that get attached to particular proteins.* Where the respective dyes show up, we know that the genes that encode those proteins have become activated. The results are a revelation (Fig. 7.3).

Stripes! If Alan Turing had known that this is what a fruit fly embryo looked like a few hours after fertilization, he could have been forgiven for thinking that he had cracked the problem of morphogenesis. For it should be clear by now that stripe patterns are one of the distinctive features of a reaction–diffusion process characterized by Turing's activation and inhibition. So isn't this proof that biological pattern begins as Turing structures?

But no, the fly's stripes do not vindicate Turing after all. For one thing, they are never Turing-style stripes, with their trademark wiggles and bifurcations, but are always regular bands running more or less perpendicular to the embryo's head-to-tail (anterior–posterior) axis. Nor are the bands equivalent: they are, as it were, merely signs of a larger plan.

Let us go back to the fertilized egg, which in humans and many other species is a sphere. To progress from this to a newborn baby, a lot of symmetry must be broken. Turing's mechanism provides a way to do that, but there is no reason to suppose that it is unique. Today's understanding of morphogenesis suggests that here, at least, nature may use tricks that are at the same time less complex and elegant but more complicated than Turing's reaction–diffusion instability. It seems that eggs are patterned and compartmentalized not by a single, global mechanism but by a sequence of rather cruder processes that achieve their goal by virtue of their multiplicity.

The reference grid of a fertilized egg, which tells cells whether they lie in the region that will become the head, a leg, a vertebra, or whatever, is apparently painted by diffusing biochemicals. But these substances merely trace out monotonous gradients of concentration, high near their source

---

*A common approach is to modify the genes of the respective proteins by genetic engineering so that, when the proteins are made, they already have a fluorescent chemical group attached. The appearance of fluorescence then signals that the gene is switched on.

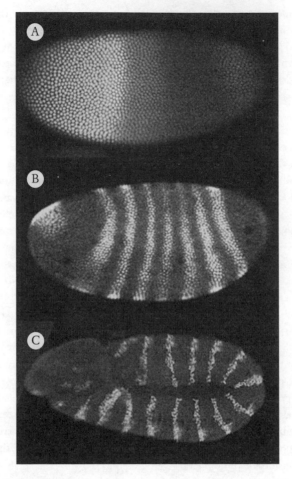

FIG. 7.3  Genes in the fruit fly embryo are 'switched on' in stripe patterns in the early stages of growth, as revealed here by coloured light emission from the proteins that the genes encode. Initially, two proteins (called hunchback and Kruppel) are generated in broad bands (a). This changes to narrow bands of another gene (b), followed by the beginnings of true segmentation (c). (Photos: Jim Langeland and Steve Paddock, kindly provided by Sean Carroll, University of Wisconsin-Madison. From Carroll, 2005.)

and decreasing with increasing distance. A gradient of this sort differentiates space, providing a directional arrow that points down the slope of the gradient. Each of the chemical morphogens has a limited potential by itself to structure the egg, but several of them, launched from different sources, are enough to get the growth process under way by providing a criss-crossing of diffusional gradients that establish top from bottom, right from left; in other words, they break the symmetry of the egg and sketch out the fundamentals of the body plan.

The idea of chemical gradients as the coordinators of the first stages of morphogenesis predates Turing—it can be traced back to the German biologist Theodor Boveri, who suggested in 1901 that changes in concentration of some chemical species from one end of the egg to the other might control development. Experiments involving the transplantation of cells in early embryos led the British biologist Julian Huxley, along with embryologist Gavin De Beer, to propose in 1934 that small groups of cells, called organizing centres or organizers, set up 'developmental fields' in the fertilized egg that are responsible for the first patterning steps. If these organizers were transplanted to different parts of the fertilized egg, they produced new patterns of subsequent development, suggesting that the organizers exercise a local influence on the cells around them.

In 1969 the British biologist Lewis Wolpert marshalled these ideas into the beginnings of the modern view of morphogenesis. He asserted that the diffusional gradients of morphogens emanating from organizing centres provide positional information, letting cells know where they are situated in the body plan. Wolpert proposed that morphogens are generated in clusters of cells that had previously been identified and called zones of polarizing activity (ZPAs).

The business of how morphogen gradients work is best understood in the fruit fly. The fruit fly egg is unusual in that it does not become compartmentalized into many cells separated by membranes until a relatively late stage in the growth process, by which time much of the essential body plan is laid down. Like all developing eggs, the fruit fly egg makes copies of its central nucleus, where the genetic storehouses of the chromosomes reside. But whereas in most organisms these replicated nuclei then become segregated into separate cells, the fruit fly egg just accumulates them around its periphery. Only when there are about 6,000 nuclei in the egg do they start to acquire their own membranes.

For this reason, morphogens in the fruit fly embryo are free to diffuse throughout the egg in the first few hours after it is laid. After a short time, the egg develops the stripes we noted earlier. These evolve into a series of finer stripes, marking out regions that will ultimately become different body segments: the head, the thorax, the abdomen, and so forth.

The first breaking of symmetry takes place along the long (anterior–posterior) axis of the egg. The initial segmentation process seems to be controlled by three genetically encoded signals: one defines the head and thorax area, another the abdomen, and a third controls the development of structures at the tips of the head and tail. When the respective genes are activated, they generate a morphogen that diffuses from the signalling site throughout the rest of the egg.

The head/thorax morphogen is a protein called bicoid. Production of the bicoid protein takes place at the extreme anterior end of the egg, and the protein diffuses through the cell to establish a smoothly declining concentration gradient (Fig. 7.4a). To transform this smooth gradient into a sharp compartmental boundary (which will subsequently define the extent of the head and thorax regions), nature exploits the kind of threshold switch that I have described earlier. Below a certain threshold concentration, bicoid has no effect on the egg, but above this threshold the protein binds to DNA and activates another gene, generating its corresponding protein product called hunchback. In this way, a smooth gradient in one molecule (bicoid) is converted into an abruptly stepped variation in another (hunchback) (Fig. 7.4b,c).

You may have noticed that this patterning mechanism seems to have cheated on the question posed at the outset: how does an initially uniform egg break its symmetry? True, the fruit-fly egg is not quite so uniform—it already has a long axis and a short axis. But why should bicoid suddenly be produced at one end and not the other, or indeed in any one region of the egg and not others? The answer seems to be that the egg is subjected to asymmetric outside interference. Although the egg itself is initially a single cell, it begins its development as a part of a multicellular body. The single 'germ cell' that will grow into the egg becomes attached to follicle cells

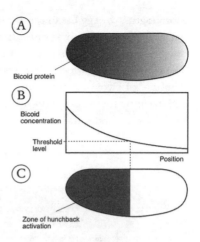

FIG. 7.4  Initial patterning in the fruit fly embryo is controlled by a protein called bicoid, which diffuses along the cell from the anterior end to set up a concentration gradient (*a*). Where the concentration surpasses a certain threshold, the bicoid protein triggers the formation of the so-called hunchback protein (*b*, *c*). Thus the smooth gradient in bicoid gives rise to an abrupt boundary of hunchback expression.

before fertilization, and within this assembly the follicle cells and other specialized entities called nurse cells provide nutrients for the egg cell's growth. The nurse cells deposit molecules* encoding the bicoid protein at the anterior tip of the egg while they are still attached to one another, and in this way the bicoid protein is generated as soon as the cell is fertilized. So there is no wondrous spontaneous symmetry-breaking here as there is in Turing's mechanism. Instead, a broken symmetry is passed from generation to generation.

The patterning of the posterior region of the fruit fly egg is controlled by a morphogen called the nanos protein. (Nanos is Greek for dwarf, and what with hunchbacks too, you can imagine that there are unfortunate deformities associated with the malfunctioning of these genes.) At some stage after longitudinal segmentation has been induced by these

---

*These are RNA molecules, which act as the intermediaries in the translation of chromosomal DNA to protein. The information on a single gene in the chromosomes is first transcribed to RNA, and the RNA molecules then serve as a template for protein synthesis by a multi-molecule assembly called the ribosome.

morphogens, the egg has to break the symmetry between top (where the wings will go) and bottom (where the legs and belly are). This is called the dorsoventral axis, and its direction is defined by a protein called dorsal. The mechanism by which dorsal does its job is rather more complicated than that of bicoid or nanos, however. The top–bottom gradient is not one in concentration of the dorsal protein—which is actually more or less uniform throughout the egg—but in the nature of the protein's position

FIG. 7.5 'Latitudinal' bands in the fruit fly embryo are created by genes that are turned on in the bottom, middle and top regions. (Photos: Michael Levine, University of California at Berkeley, kindly provided by Sean Carroll, University of Wisconsin-Madison. From Carroll, 2005.)

within individual cells. Towards the bottom, it segregates more strongly into nuclei than into the cell's watery cytoplasm, while the reverse is true towards the top. There appears to be an underlying signal of still uncertain nature that determines whether or not the dorsal protein can find its way into the many nuclei in the egg; this signal is activated from the bottom (ventral) edge of the embryo (Fig. 7.5). Again, the initial impulse for this symmetry-breaking signal seems to come from outside the cell—from a concentration gradient in some protein diffusing through the extracellular medium, which transmits its presence to the egg's interior by interactions at the cell membrane. You could say that in this sense nature cheats somewhat in breaking the initial symmetry of the egg.

## OPENING THE HOMEOBOX

All this might make it seem as though morphogenesis, far from being a process of spontaneous patterning, is instead just one thing after another. In a sense that is true; but this does not mean that it is merely arbitrary, its shapes being defined simply by the turning of genetic dials and throwing of genetic switches. These shapes, as we have seen, are not like portraits drawn line by line in each patch of the canvas until the picture is complete. They are instead the result of a process of sequential elaboration caused by the superposition of *global* patterning signals. It is in this, admittedly restrictive, sense that the human body can be regarded as a kind of genuine pattern—one that is tightly controlled, but based initially on rather simple symmetries.

Say, for example, that we wish to generate a specific and small number of spots of cells of a certain type on the middle of the rear body of the fruit fly. This is just what happens to initiate formation of the limbs: the legs and wings. We have already seen that the embryo gets divided into stripes. Now, morphogens coming from the top and bottom can inactivate certain genes in those bands everywhere except along a stripe in the middle of the body. Then a similar switching process initiated from the head confines the resulting spots to the rear half of the body (7.6a). Something just like that occurs to create the regions that will bud eventually into legs (Fig. 7.6b).

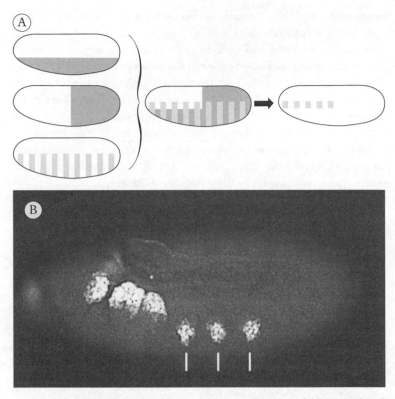

FIG. 7.6 How to make spots from stripes. A combination of longitudinal and latitudinal bands of protein expression creates discrete patches where the bands overlap—here, in the anterior equatorial region (*a*). Such spots can be seen, for example, in genes that initiate the formation of legs (the three lower bright regions in *b*, marked by 'l'.) (Photo: *b*, Scott Weatherbee, kindly provided by Sean Carroll, University of Wisconsin-Madison. From Carroll, 2005.)

You can see that this is rather like the toolbox idea invoked earlier to account for the range of pattern features on butterfly wings. What you would not expect, however, is that this is not merely analogy: for the eyespot of the butterfly wing and the legs of the fruit fly are triggered by the very same gene, *Distal-less*! I touched on this in Chapter 4, but it is actually downright bizarre. Why should a gene that produces a wing-scale pigmentation pattern in butterflies have anything to do with one that makes legs in flies?

In the answer to this question lies the second reason why I believe we are justified in regarding morphogenetic patterning as 'universal'—if not in the same way that Turing patterns are universal, then at least in so far as it invokes processes that are common to all living creatures. For it has been discovered over the past two decades that morphogenesis is controlled by a relatively small number of genes that are shared across diverse species, from flies to humans. Not only may these genes serve very different roles in different creatures, as is the case with *Distal-less*, but they may perform several quite distinct functions in single organisms, depending on when and where they are switched on. These genes are controlled by the morphogen gradients that divide developing embryos into grids and specify the fate of each cell within that grid.

A clue to the existence of genes responsible for very general features of an organism's body plan is provided by grotesque mutations in which body parts grow in the wrong position. One such mutant form of the fruit fly was identified in 1915 by the biologist Calvin Bridges at Columbia University in New York. He found and selectively bred flies in which the hindwings, which are normally small, grow to resemble the large forewings. A more striking later example was a fly in which legs grow in place of antennae on the head. In 1894 the zoologist Gregory Bateson called such mutants, in which one body part grows to resemble another, *homeotic*. An antenna that becomes a leg seems very strange, because one might reasonably think that the growth of each of these appendages is guided by independent suites of genes. If that's the case, why should they all start acting erratically together?

But geneticists discovered that such homeotic mutations seem to be linked to just a single gene—a so-called homeotic gene. Even more surprisingly, when in the 1980s it became possible to examine the DNA sequences of homeotic genes responsible for various mutations of fruit flies, it was found that all of them contained a short stretch of DNA that was identical. This segment was named the homeobox, and genes containing it were dubbed homeobox or *Hox* genes.

*Hox* genes have been identified in humans and other mammals too, and they share almost identical homeobox DNA sequences with those in fruit flies. This implies that the homeobox is a rather ancient genetic element that determines body plans. Particular *Hox* genes, and other 'patterning' genes with domains closely related to the homeobox,* seem to be associated with specific body-building functions in very different species. The *eyeless* gene in flies, mutations of which cause a loss of eyes, is equivalent to (the technical term is 'homologous with') a gene called *Aniridia* in humans, which reduces or eliminates the iris, and with a self-explanatory gene called *small eye* in mice. When geneticists activated the *eyeless* gene in other tissues of flies, it induced the formation of rudimentary eyes. Still more strikingly, transplanting the *small eye* gene of mice into such fly tissues also initiated eye formation—but of compound flies' eyes, not mouse eyes. It is tempting to say that all these homologous genes, known collectively by the family name *Pax-6*, apparently encode the instruction 'start making the eye appropriate for your species'. Yet *Pax-6* plays several other roles in mammalian development too, being involved in (among other things) the development of the brain, nose, gut, and pancreas. That, as already mentioned, is a characteristic of these patterning genes: they do not encode any specific developmental fate, but are mere *routers*, like train-line signals, the effects of which depend on precisely when and where in the developmental process they get switched on. In each case, once that signal is given, it directs development down a particular track and other genes responsible for the details (for example, for making each of the components of an eye) come into play further on down the line.

Just as one of the generic roles of *Pax-6* is making eyes, *Distal-less* may act as a general-purpose 'appendage-inducing' gene. In insects it produces parts of limbs, in chickens it makes legs, in fish fins, and in sea squirts the squashy bulbs called ampullae. Yet, remarkably, butterflies have commandeered this gene for the completely different patterning function of making eyespot markings. So, again, such genes cannot be regarded as having anything 'limb-like' about them, as it were—they are mere switches that happen often to be incorporated into appendage-making genetic circuits. This multiplicity of uses for homeotic genes is reflected in the fact that they typically come equipped with several 'controls' that enable them to respond to different kinds of morphogenetic signal. This permits them to display a rather complex logic: for example, some might be activated only if both of two morphogens are present. By this means,

---

*Some 24 distinct families of such genetic segments, called homeodomains, are now known.

rather fine patterning features can be defined: they might be created by the overlap of several morphogen fields, say. Homeotic mutants arise when the genetic switches break or malfunction.

The fates of cells are marked out in an embryo very early in development: experiments have shown that long before embryos show any sign of growing, say, legs or wings or eyes, certain cells have been earmarked for that function by the activation of the requisite suites of genes. Geneticists have constructed 'fate maps' which show the destinies of different regions of an embryo. The body-patterning process then becomes a question of defining the coordinates of these maps, and activating the right homeotic genes in the right places at the right time. These tasks are apparently conducted by diffusing molecules such as the bicoid protein, issuing from locations defined in earlier patterning steps, right back to the initial symmetry-breaking of the cluster of cells. In this way, crude grids are progressively converted into ever finely marked and differentiated ones.

The gridding process has been followed in some detail in fruit fly morphogenesis. A few hours after fertilization, the embryo is about 100 cells long. It gets divided first into bands at one end and in the middle (which overlap slightly) by the expression of the hunchback protein, and another called Kruppel. Then these are divided into precisely seven stripes, each three to four cells wide, separated by 'interstripes' of similar width (Fig. 7.3a,b). Crucially, in relation to the stripy patterns generated from Turing-like mechanisms, the stripes here are not all equivalent: each of them is defined by a different setting of genetic switches, produced by the distinct pattern of morphogenetic signals at each position. Turing's scheme offers a simpler way to make what appears to be the same kind of pattern, but apparently it does not imprint enough information on the system to ensure the subsequent evolution of the pattern.

Gradually these stripes fade and are replaced by others along both the anterior–posterior and dorsoventral axes. This is because each set of 'stripe proteins' turns on genes that produce a subsequent set, with a new pattern that elaborates the previous one: each pattern contains the seed of the next. The intersection of these 'latitudinal' and longitudinal' bands introduces two-dimensional patterning (Fig. 7.7) in a cascade of increasing complexity—and we can see the schematic plan of a fly start to emerge.

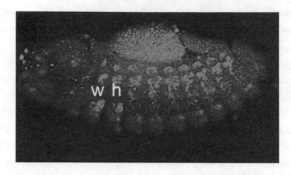

FIG. 7.7 After several rounds of banded patterning, the body plan of the fly begins to emerge. Here two of the bright spots, marked 'w' and 'h', denote the positions of the future forewing and hindwing. (Photo: Scott Weatherbee, kindly provided by Sean Carroll, University of Wisconsin-Madison. From Carroll, 2005.)

## HAIR TODAY

Does all this mean that Turing's bold hypothesis has nothing to do with real morphogenesis, beyond perhaps the formation of animal markings? Let us set aside for a moment the question of patterning as such, and think instead simply of Turing's notion of a diffusing chemical that acts as a morphogen. As we have seen, this idea was in fact inherent in the earlier work, but it was Turing who first made the proposal explicit and showed how it could create rather elaborate patterns. Yet is there anything in it?

There is now good reason to think so. In 1990, researchers at the National Institute for Medical Research in Mill Hill, London, showed that a protein called Activin appears to act as a morphogen in the African clawed toad, determining the fate of embryonic cells called mesoderm cells. At a high concentration, Activin induces mesoderm cells to develop into a primitive backbone called the notochord, whereas at low concentrations of the protein these cells become muscle tissue. Soon after this discovery, several groups of researchers found evidence that various proteins act as morphogens in fruit flies. When a morphogen arrives at a cell, it binds to a molecule called a receptor at the cell surface, triggering a sequence of biochemical events inside the cell that set it on a particular developmental pathway.

It is hard in such cases to be sure that the putative morphogen really is diffusing throughout the embryo, like a travelling salesman knocking on the door of each cell. One alternative is that the patterning signal moves in a kind of relay: each cell, activated by a morphogen arriving at its surface, sends out a signal to its neighbours. In 1994, however, a team led by John Gurdon at the Wellcome Cancer Research Centre in Cambridge showed that Activin in frog eggs really does seem to get about in the space between cells by simple diffusion, like ink dispersing in water. And seven years later, Yu Chen and Alexander Schier of the New York School of Medicine showed that diffusion, rather than a relay mechanism, explains how a protein called Squint determines the fates of mesoderm cells in the embryonic zebrafish. Activin and Squint are both members of a family of related proteins called transforming growth factor $\beta$ (TGF$\beta$) proteins, which are of central importance in embryogenesis. There seems good reason to think, then, that some of these proteins deliver patterning instructions by wandering over long distances.

Morphogens do not always act this way, however. Take the protein Decapentaplegic (Dpp), another type of TGF$\beta$ protein. Its local concentration in fruit fly embryos was found in 1996 to 'tell' cells where they are along the front–back axis of wings, thereby switching on certain genes that induce the cells to grow into different tissues. It looks as though Dpp is acting here as another classic morphogen. But experiments in 2000 by Marcos González-Gaitán and his colleagues at the Max-Planck Institute for Molecular Cell Biology and Genetics in Dresden showed that Dpp seems to be captured by proteins at cell surfaces and carried inside—the salesman, knocking at the door, not only gets a response but is invited in. So, rather than wandering randomly among cells—that is, genuinely diffusing—the morphogen relies on molecular guides to usher it across cell membranes. Thus, the cells are not passive puppets of the morphogen gradient; instead they actively shape that gradient. What is more, Dpp exerts its effects in another way too: exposure of a cell to a local dose of Dpp at one stage of development can programme it to develop a certain way at a later stage, when the cell was far from the Dpp source. In other words, cells may retain a memory of contact with Dpp, so that the local concentration of this protein may be irrelevant when the cells start to respond to that prior 'programming'. This all shows that the concentration gradient of a freely wandering morphogen is not the only way for it to encode positional information.

What Turing had in mind, however, was not simply a suite of messenger molecules telling cells to do this or that. His patterns came about through the competition of an activating and an inhibiting process. This was a 'global' phenomenon, rather than being governed by localized events that initiate particular developmental pathways at specific places. Diffusing signals of activation and inhibition produce a field of regular, roughly identical and *self-organized* features. Is there any space for that sort of thing in morphogenesis?

There now appears to be at least one general type of embryonic patterning process, apart from pigmentation, that bears some relation to Turing's scheme. Many tissues and components of living organisms seem to exhibit patterns in the sense I have generally used: the more or less regular repetition of identical elements. We see it, for instance, in the distribution of hairs on skin—look at the back of your hand, and you will see that not only are the follicles roughly equidistant, but there is the suggestion of hexagonal packing among them (Fig. 7.8a). The same is true for the spacing of feathers in birds, while the scales of lizards and butterfly wings have much more pronounced orderliness in their arrangement (Fig. 7.8b,c). In chicks the feather buds are marked out by a gene called *patched* (which has other patterning roles in other organisms: it is involved in the segmentation of fruit flies, for instance). It seems that Turing's concept of *inhibition* plays a central role in creating the regularity of pattern features. As cells in a homogeneous mass, such as those of the nascent skin, begin to differentiate to form the specialized structures and tissues that make feathers, they send out diffusing molecular signals to the cells nearby that suppress such differentiation, ensuring that two such patches do not develop close together. There is thus local activation in the sense that cells can become spontaneously committed to differentiation into feather-making apparatus, and longer-ranged inhibition of the same thing happening in the vicinity.

Something similar applies to the formation of hair follicles in mice. Here a protein called Wnt appears to function as an activator of follicle formation, while a class of proteins known generically as Dkk acts as inhibitors of Wnt. Thomas Schlake and his co-workers at the University of Freiburg in Germany showed in 2006 that genetically mutant mice that produced Dkk proteins in abnormally high amounts developed follicle

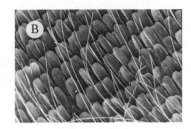

FIG. 7.8 Regular spot patterns in biology: human hair follicles (*a*), butterfly scales (*b*) and snake scales (*c*). (Photos: *a*, Martin Dohrn / Science Photo Library; *b*, Centre for Electon Optical Studies, University of Bath; *c*, Mohammed Al-Naser.)

patterns that matched with those predicted theoretically from Turing-style activator–inhibitor models of the diffusion and interaction of Wnt and Dkk.

So while Turing's patterns are perhaps too imprecise and uniform to be of much general help in building an organism, it seems that they are just the job when some regular, repeating pattern is required. The key point in such cases is that the pattern features in these structures do not need to be precisely defined on any global coordinate grid (of the kind that may be 'drawn' by morphogen gradients); it is enough that they are merely placed at roughly even spacing. This means that the features can be determined simply by *local* rules, applied globally: by interactions between pattern features played out within an 'active' matrix filled with pattern-forming potential. *That* is self-made patterning (what scientists now tend to call emergence) in its truest sense.

## THE SHAPE OF THINGS TO COME

One of the most satisfying aspects of the story of homeotic genes is that it achieves at last the synthesis of evolution and embryogenesis—of phylogeny and ontogeny—that Ernst Haeckel sought. This reconciliation, dubbed 'evo devo' (the melding of evolution and development), is based on the notion of a toolbox of patterning genes that, as well as being activated at different times in the growth of the embryo, have been enlisted to serve different evolutionary functions. Clearly, butterflies did not need a *new* gene to produce eyespots—that role could be fulfilled by a gene already present in the ancestral organism, although originally it did nothing of the sort.

This new perspective on morphogenesis helps us to make sense of some of the episodes in evolutionary time when there seems to have been a sudden diversification of body plans. Most notably, the explosion of animal varieties that took place at the start of the Cambrian period (Fig. 7.9), around 540 million years ago, can now be seen as the consequence not of some mysterious acquisition of lots of new genes, but of the appearance of mechanisms—genetic switches—for programming existing genes in new ways. The living world already had the tools before diversification happened; the explosion of variety was the result of nature working out how to use them. One can draw a crude analogy with musical or literary creativity: when the penny drops that the rules of grammar permit so much more than the construction of single sentences, you can begin to write stories.

It is here that the neo-Darwinist Modern Synthesis falters. Each living creature, it seems, is not a story to itself, but an elaboration of a handful of grand narratives that contain all possible stories, some of them connected in the most surprising ways. Or perhaps to sharpen that analogy: the grammar and syntax of genetics transcend individual species.

This, however, does not undermine but only enriches the picture of the evolutionary process that has emerged since Darwin. Indeed, it does so in a way that silences some of the objections to evolution that fundamentalist religious critics have raised. New functions—vision, say, or flight—did not have to be 'invented from scratch' using a whole new set of genes; instead, small changes in the way existing genes were regulated and activated gave rise to structures that were modifications of old ones and yet at the same time had entirely new functions. Nature does not have, or

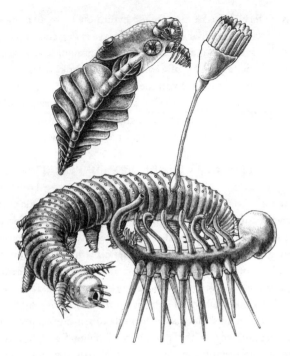

FIG. 7.9 The Cambrian period was a time of tremendous experimentation in body plans. Here are just a few of the bizarre creatures reconstructed from fossilized remains found in the Burgess Shale in the Canadian Rockies. Clockwise from top left: *Anomalocaris, Aysheaia, Hallucigenia* and *Dinomischus*. (After Marianne Collins.)

need, the foresight of the engineer: she tinkers blindly with what is to hand, and yet is fortunate to have hit on a way of patterning embryos that is extraordinarily fecund, offering profound change from minor adjustments. There are, of course, drawbacks to this way of making organisms, the clearest of which is that the breakdown of one patterning step can have serious or even catastrophic effects further down the line. If the initial error goes unnoticed, the subsequent steps are played out blindly in an entirely inappropriate context. That is how so-called 'monsters' (creatures with dramatically altered morphologies) arise. In this finely poised series of steps, one slip can derail the whole process, creating growth anomalies that are more often than not fatal.

This is a very different picture of evolution from the one D'Arcy Thompson drew by bending grids—for it is one thing to see that a baboon skull might be mutated (graphically, not biologically!) into a human one, but the genetic patterning kit of evo devo reveals links between creatures that would never be guessed from their body shapes. More generally, evo devo and the genetics of form offer a means of exploring evolutionary questions that do not depend on the rather subjective issue of similarities of form, which has guided comparative anatomy in the past.

## THE INEVITABILITY OF FORM?

The existence of *Hox* genes goes to the heart of the question that both Ernst Haeckel and D'Arcy Thompson confronted in their different ways. We have seen that Haeckel had a rather Hegelian view of evolution in which living form obeys imperatives shaped by ill-defined forces, one of which is a kind of Platonic tendency towards symmetry. Thompson felt that the apparently blank sheet offered by Darwinists is in fact constrained considerably by the mechanical exigencies of physical law, making certain structures more or less inevitable. In both cases, the suggestion is of a degree of determinacy in the evolution of living form. If this form is controlled by a rather small number of genes, does this mean that there is only a finite number of possible outcomes dictated by their various permutations applied to a spherical ball of cells? In other words, were Thompson and Haeckel right after all to impose a finite universe of pattern and form?

Possibly so—but the number of permutations is astronomical. Certainly, the ways in which known *Hox* genes are regulated and switched are far more numerous than evolution could possibly have explored.* Our

---

*When the *Hox* genes were first identified, evolutionary biologists assumed that the increasing complexity of body patterns throughout evolutionary history was the result of their acquiring more of these 'switches' by simply making copies in their genomes. But in fact even very 'primitive' organisms have been shown to have impressive arrays of *Hox* genes, suggesting that complexity of form arose not from having more patterning switches but from finding new ways of controlling and configuring them. By working out what these genes were in ancestral organisms, now long since extinct, it is possible to make good guesses about what these organisms may have looked like—whether, for example, the last common ancestor of humans and flies had eyes or a central nervous system.

world can only have hosted a small fraction of this galaxy of Platonic organisms. Does this mean that, if we were to run evolution over again, the results might be quite different?

That is one of biologists' favourite questions (which is a way of saying that they argue over it furiously). The instinct of the whole of biology runs counter to that which has characterized the physical sciences, in that it opposes any notion of determinacy. For neo-Darwinists, randomness is the order of the day, pruned by the struggle to survive. Anything else smells to some biologists of creationism by stealth—which is to say, that loathed pseudoscience 'intelligent design'. That is a shame, for the question is a real one, and there is much fun as well as serious science to be found in exploring the answer.

For one thing, it is well known now that selective pressure can channel evolution towards a particular morphological 'solution' along independent routes: this is the phenomenon known as convergent evolution. The shapes of fish bodies, of insect appendages, of carnivorous mammal teeth, of bird wing structures, of plant bodies, all show similarities that may come not from shared descent but from unconnected 'discoveries' of what we might call a good engineering design during the independent evolution of the species concerned. Traditionally, these coincidences have not been deemed to be saying anything about the breadth of nature's palette, for convergent evolution seems to suggest merely that in certain fixed circumstances there is a 'best', or at least a preferable, way to survive: that commonalities of form 'may be imposed from without, not from the basic biology of morphogenesis.

But the discovery of patterning genes such as *Pax-6* and *Distal-less* has sparked a heated discussion about the significance of convergent evolution. Does the similar joined character of insect and human limbs, say, indicate that these are the 'best' solutions and that nature has sought them out as such, or does it mean that they are simply the solutions that the available set of patterning genes permit? This is not at all easy to answer, not least because we really have no idea what the intrinsic limitations of the set of patterning genes, coupled to the rest of the genome in an incredibly complex network of interactions that plays itself out over time, actually are. These are not, after all, genes 'for' making legs, or eyes, or whatever—they are routers with multiple functions. Thus the evolutionary biologist Simon Conway Morris points out that it seems quite permissible, at least, to regard these genes as mere hitchhikers carried along the paths of convergent evolution. The ancestral precursor of the *Distal-less* gene, for example, may have been linked to the development of sensory organs; and so its involvement in limb growth may not say anything about the particular options available to the *Distal-less* patterning pathway when embedded in the gene network, but may simply be a consequence of the fact that sensory organs were often placed on protrusions from the body—so that *Distal-less* was a convenient switch for controlling the formation of appendages.

## THE POSSIBLE AND THE ACTUAL

As D'Arcy Thompson showed, there are many arguments suggesting why certain biological forms are likely to be dictated by the laws of physics as played out in their particular environment. If bubbles and foams are used to create reticulated networks, we can expect them to display the geometries that Plateau elucidated. If reaction–diffusion processes are exploited to create markings and other patterns, we can anticipate Turing's spots and stripes and the various stunning elaborations thereof. In that much, we might argue for a degree of universality in the patterning of the living world. Does this generality extend to the more complex forms of bodies and anatomies? We cannot say, although we can be sure that the palette is at the very least far more diverse.

But of this much we can be more certain: the notion of Haeckel and the *Naturphilosophen* that an organizing force of nature lies at the root of all morphogenesis is very far from finding support in modern biological science. We can make a convincing case, and I hope I have done so here, that living nature exploits pattern formation in many ways. But it seems to do so essentially as an opportunist, not as a minion dictated by any Higher Force.

## SOAP-FILM STRUCTURES

I t is well worth investigating the pleasing symmetrical figures produced by Plateau's rules for intersecting soap films when suspended
on wire frames. These are fairly easy to make from copper wire even
with a soldering technique as crude as mine. Some of the simplest
possibilities are shown in Fig. 2.24. I have used 1-mm copper wire to
construct polygonal frames with sides about 5 cm long. The task is
made easier by bending a continuous piece of wire to construct as much
of the frame as you can, rather than trying to join all the struts separately.
Remember to include an extra appendage as a handle.

These frames can be dipped into a solution of washing-up liquid. These
solutions lose their film-making ability over time. For more complex
frames, such as the octahedron, the resulting pattern of intersecting film
faces is not unique, and you can induce rearrangements by blowing
gently. If small extraneous bubbles get trapped along some vertices,
these can be removed by pricking them carefully with a pin.

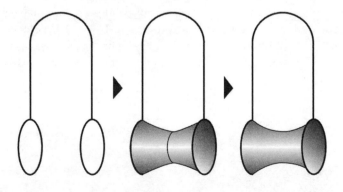

To make a catenoid (Fig. 2.32), you can construct a tweezer-like frame with two circular loops at the ends. Dipping this frame will often produce two half-catenoids with a circular film between them; this can be converted to a catenoid by pricking the central film. You'll find that the films are surprisingly resilient when freshly made.

For a wider range of experiments with bubbles and foams, I highly recommend C. V. Boys' classic text *Soap Bubbles* (1959).

## OSCILLATING CHEMICAL REACTIONS

T here is a variety of reliable oscillatory chemical reactions described in the chemistry literature, including several recipes in books intended for teaching or for a non-specialist readership (see below). One of the most striking colour changes is provided by the so-called iodate/iodine/peroxide oscillator. The recipe that I have used is as follows:

> Solution A: 200 ml of potassium iodate ($KIO_3$) solution, made by adding 42.8g of $KIO_3$ and 80 ml of 2M sulphuric acid to distilled water to make a total volume of 1 litre.
>
> Solution B: 200 ml of malonic acid/manganese sulphate ($MnSO_4$) solution, made by adding 15.6g malonic acid and 4.45g $MnSO_4$ to distilled water to a total of 1 litre.
>
> Solution C: 40 ml of 1% starch solution, made by adding a slurry of 'soluble' starch to boiling water.
>
> Solution D: 200 ml of 100 vol. (about 30%) hydrogen peroxide ($H_2O_2$) solution.

Mix solutions A, B and C together in a conical flask and then initiate the reaction by adding solution D. Mix well using a magnetic stirrer. After a minute or two the solution, which is initially blue (owing to the formation of iodine, which reacts with starch to form a blue compound), turns a pale yellow (as the iodine disappears), and then abruptly blue again as another cycle begins. The colour changes persist for about 15–20 minutes, but finally they run out of steam because some of the initial reagents are used up in each cycle.

After a few minutes the mixture begins to bubble, as carbon dioxide gas is generated from the oxidation of malonic acid.

If the mixture is not stirred, the colour changes still take place but they grow from filamentary patches throughout the solution.

It is important not to prepare the malonic acid solution too far in advance, because it begins to decompose over the course of several weeks.

The most famous oscillatory reaction is the Belousov–Zhabotinsky reaction. Here is the recipe I have used for this:

> Solution A: 400 ml of 0.5M malonic acid solution, made from 52.1g malonic acid dissolved in 1 litre of water.
> Solution B: 200 ml of 0.01M cerium (IV) sulphate ($Ce(SO_4)_2$) in 6M sulphuric acid.
> Solution C: 0.25M potassium bromate ($KBrO_3$), made from 41.8g $KBrO_3$ in 1 litre of water.

Mix solutions A and B in a magnetically stirred conical flask, and then add solution C to start the reaction. After about 3 minutes, the solution starts to alternate between colourless and yellow. The oscillations last for 10–15 minutes.

This is the colour change that Belousov first saw; but it can be made more dramatic by adding 1 ml of an indicator called ferroin (iron tris(phenanthroline)), which makes the solution change between blue and purplish red.

I have taken these recipes from the chemical demonstrations leaflet of the chemistry department of University College, London, and am very grateful to Graeme Hogarth and Andrea Sella for help in carrying out these experiments and those in the following two appendices. Other oscillating reactions, and variations of these two recipes, can be found in:

Shakhashiri, B. Z., *Chemical Demonstrations: A Handbook for Teachers of Chemistry*, vols 2 and 4 (Madison: University of Wisconsin Press, 1992).

Roesky, H. W., and Möckel, K., *Chemical Curiosities* (Weinheim: VCH, 1996).

Ford, L. A., *Chemical Magic* (New York: Dover, 1993).

## CHEMICAL WAVES IN THE BZ REACTION

T he target patterns of the unstirred Belousov-Zhabotinsky (BZ) reaction always looked to me so extraordinary that I found it hard to believe they would be easy to make. But they are, although it's another matter to create the perfect circular rings commonly seen in textbook images (those generally require the diffusion of the molecules to be slowed down by conducting the reaction in a gel rather than in water). This recipe seems very reliable:

Solution A: 2 ml sulphuric acid and 5g sodium bromate (NaBrO$_3$) in 67 ml water.
Solution B: 1g sodium bromide (NaBr) in 10 ml water.
Solution C: 1g malonic acid in 10 ml water.
Solution D: 1 ml ferroin (25 mM phenanthroline ferrous sulphate).
Solution E: 1g Triton X-100 (a kind of detergent) in 1 litre of water.

Put 6 ml of solution A into a Petri dish about 3 inches in diameter, add 1–2 ml of solution B and 1 ml of solution C. The solution turns a brownish colour as bromine is produced. Make sure you don't inhale deeply over the dish—bromine is noxious. After a minute or so the brown colour will disappear. Once the solution has become clear, add 1 ml of solution D (which will turn the liquid red) and a drop of solution E. Swirl the Petri dish gently to mix the solutions—it will turn blue as you do so, but then quickly reverts to red. Then leave it to stand. Gradually, blue spots will appear in the quiescent red liquid, and these will slowly expand as circular wave fronts. Typically there will be one to a dozen or so separate target-wave centres, and the blue wavefronts annihilate one another as they collide.

This reaction can be shown to an audience by placing the dish on an overhead projector. The heat of the projector will warm the solution and speed up the motion of the wavefronts a little. After some time, bubbles of carbon dioxide will start to appear. These can obscure or disrupt the pattern, but you can get rid of them and restart the process by swirling the solution gently.

This recipe is taken from the chemical demonstrations leaflet of the chemistry department of University College, London.

## LIESEGANG BANDS

T his is a lovely experiment, but it takes several days. The bands are zones of precipitation of an insoluble compound, which occurs at intervals down a column filled with a gel, through which one of the reagents diffuses from the top.

You can use a burette as the column (about 1 cm diameter), although ideally a glass tube without gradation markings gives the clearest view. The recipe I have used involves the reaction between cobalt chloride and ammonium hydroxide, which precipitates bluish bands of cobalt hydroxide. The cobalt chloride is dispersed in a gelatin gel: mix 1.5g of fine-grained gelatin and 1g of hydrous cobalt chloride ($CoCl_2.6H_2O$) with 25 ml of distilled water and heat to boiling point for five minutes. Then transfer this mixture immediately to the glass column, cover the top of the column with plastic film, and allow it to stand for 24 hours at room temperature (about 22 °C) for the gel to set.

Then add 1.5 ml of concentrated ammonia solution to the top of the solidified gel using a pipette. Cover the tube again and leave it to stand.

After several days, the bands begin to appear down the column. They are closely spaced—about a millimetre apart, although the spacing is not constant (see Chapter 3). You have to get on eye level with the bands to see them clearly, but they should be sharp and well defined.

This recipe is taken from:

Sultan, R., and Sadek, S., 'Patterning trends and chaotic behaviour in $Co^{2+}/NH_4OH$ Liesegang systems', *Journal of Physical Chemistry* 100 (1996): 16912.

References to other systems are given in Henisch, H. K., *Crystals in Gels and Liesegang Rings* (Cambridge: Cambridge University Press, 1988).

Agladze, K., Keener, J. P., Müller, S. C., and Panfilov, A. 'Rotating spiral waves created by geometry', *Science* 264 (1994): 1746.

Alexander, V. N., 'Neutral evolution and aesthetics: Vladimir Nabokov and insect mimicry', *Santa Fe Institute Working Paper* (2001), <http://www.santafe.edu/research/publications/wpabstract/200110057>.

Aubert, J. H., Kraynik, A. M., and Rand, P. B., 'Aqueous foams', *Scientific American* 254(5) (1986): 58.

Bascompte, J., and Solé, R. V., 'Habitat fragmentation and extinction thresholds in spatially explicit models', *Journal of Animal Ecology* 65 (1996): 465.

Ben-Jacob, E., Cohen, I., Shochet, O., Aranson, I., Levine, H., and Tsimring, L., 'Complex bacterial patterns', *Nature* 373 (1995): 566.

Boissonade, J., Dulos, E., and De Kepper, P., 'Turing patterns: from myth to reality', in Kapral, R., and Showalter, K. (eds), *Chemical Waves and Patterns* (Dordrecht: Kluwer Academic, 1995).

Bonabeau, E., 'From classical models of morphogenesis to agent-based models of pattern formation', *Artificial Life* 3 (1997): 191.

Bonabeau, E., Theraulaz, G., Deneubourg, J.-L., Franks, N. R., Rafelsberger, O., Joly J.-L., and Blanco, S., 'A model for the emergence of pillars, walls and royal chambers in termite nests', *Philosophical Transactions of the Royal Society of London B* 353 (1998): 1561.

Boys, C. V., *Soap Bubbles* (New York: Dover, 1959).

Brakefield, P. M., *et al.,* 'Development, plasticity and evolution of butterfly eyespot patterns', *Nature* 384 (1996): 236.

Breidbach, O., *Visions of Nature. The Art and Science of Ernst Haeckel* (Munich: Prestel, 2006).

Brenner, M. P., Levitov, L. S., and Budrene, E. O., 'Physical mechanisms for chemotactic pattern formation in bacteria', *Biophysical Journal* 74 (1998): 1677.

Brunetti, C. R., Selegue, J. E., Monteiro, A., French, V., Brakefield, P. M., and Carroll S. B., 'The generation and diversification of butterfly eyespot color patterns', *Current Biology* 11 (2001): 1578.

Bub, G., Shrier, A., and Glass, L., 'Spiral wave generation in heterogeneous excitable media', *Physical Review Letters* 88 (2002): 058101.

Buckley, P. A., and Buckley, F. G., 'Hexagonal packing of royal tern nests', *The Auk* 94 (1977): 36.

Budrene, E. O., and Berg, H., 'Dynamics of formation of symmetrical patterns by chemotactic bacteria', *Nature* 376 (1995): 49.

Camazine, S., *Self-Organized Biological Superstructures* (Princeton: Princeton University Press, 1998).

Camazine, S., Deneubourg, J.-L., Franks, N. R., Sneyd, J., Theraulaz, G., and Bonabeau, E., *Self-Organization in Biological Systems* (Princeton: Princeton University Press, 2001).

Campos, P. R. A., de Oliveira, V. M., Giro, R., and Galvão, D. S., 'Emergence of prime numbers as a result of evolutionary strategy', *Physical Review Letters* 93 (2004): 098107.

Carroll, S. B., *Endless Forms Most Beautiful* (London: Weidenfeld & Nicolson, 2006).

Castets, V., Dulos, E., Boissonade, J., and De Kepper, P., 'Experimental evidence of a sustained standing Turing-type nonequilibrium chemical pattern', *Physical Review Letters* 64 (1990): 2953.

Chen, Y., and Schier, A. F., 'The zebrafish Nodal signal Squint functions as a morphogen', *Nature* 411 (2001): 607.

Chopard, B., Luthi, P., and Droz, M., 'Reaction-diffusion cellular automata model for the formation of Liesegang patterns', *Physical Review Letters* 72 (1994): 1384.

Church, A. H., *On the Relation of Phyllotaxis to Mechanical Laws* (London: Williams & Norgate, 1904).

Cohen, J., and Stewart, I., *The Collapse of Chaos* (London: Penguin 1994).

Conway Morris, S., *Life's Solution* (Cambridge: Cambridge University Press, 2003).

Cooke, T. J., 'Do Fibonacci numbers reveal the involvement of geometrical imperatives or biological interactions in phyllotaxis?', *Botanical Journal of the Linnean Society* 150 (2006): 3.

Copeland, B. J. (ed.), *The Essential Turing* (Oxford: Clarendon Press, 2004).

Coveney, P., and Highfield, R., *Frontiers of Complexity* (London: Faber & Faber, 1995).

Dawkins, R., *The Selfish Gene* (Oxford: Oxford University Press, 1990).

Dawkins, R., *The Blind Watchmaker* (New York: W. W. Norton, 1996).

Douady, S., and Couder Y., 'Phyllotaxis as a physical self-organized growth process', *Physical Review Letters* 68 (1992): 2098.

Dumollard, R., Carroll, J., Dupont, G., and Sardet, C., 'Calcium wave pacemakers in eggs', *Journal of Cell Science* 115 (2002): 3557.

Durian, D. J., Bideaud, H., Duringer, P., Schröder, A., Thalmann, F., and Marques, C. M., 'What is in a pebble shape?', *Physical Review Letters* 97 (2006): 028001.

Durian, D. J., Bideaud, H., Duringer, P., Schröder, A., and Marques, C. M., 'Shape and erosion of pebbles', *Physical Review E* 75 (2007): 021301.

Emmer, M., *Bolle di Sapone* (Florence: La Nuova Italia, 1991).

Entchev, E. V., Schwabedissen, A., and González-Gaitán, M., 'Gradient Formation of the TGF-β Homolog Dpp', *Cell* 103 (2000): 981.

Epstein, I. R., and Showalter, K., 'Nonlinear chemical dynamics: oscillations, patterns, and chaos', *Journal of Physical Chemistry* 100 (1996): 13132.

Ertl, G., 'Oscillatory kinetics and spatio-temporal self-organization in reactions at solid surfaces', *Science* 254 (1991): 1750.

Erwin, D. H., 'The Developmental Origins of animal body plans', in Xiao, S. and Kaufman, A. J. (eds), *Neoproterozoic Geobiology and Paleobiology* (Berlin: Springer, 2006).

Fourcade, B., Mutz, M., and Bensimon, D., 'Experimental and theoretical study of toroidal vesicles', *Physical Review Letters* 68 (1992): 2551.

Ghiradella, H., and Radigan, W., 'Development of butterfly scales: II. Struts, lattices and surface tension', *Journal of Morphology* 150 (1976): 279.

Ghiradella, H., and Radigan, W., 'Structure of butterfly scales: patterning in an insect cuticle', *Microscopy Research and Technique* 27 (1994): 429.

Ghyka, M., *The Geometry and Art of Life* (New York: Dover, 1977).

Gierer, A., and Meinhardt, H., 'A theory of biological pattern formation', *Kybernetik* 12 (1972): 30.

Gilad, E., Hardenberg, J. von, Provenzale, A., Shackak, M., and Meron, E., 'Ecosystem engineers: from pattern formation to habitat creation', *Physical Review Letters* 93 (2004): 098105.

Glass, L., 'Dynamics of cardiac arrhythmias', *Physics Today* (August 1996): 40.

Glass, L., 'Multistable spatiotemporal patterns of cardiac activity', *Proceedings of the National Academy of Sciences USA* 102 (2005): 10409.

Goethe, J. W. von, *The Collected Works*. Vol. 12: *Scientific Studies*, ed. and trans. Miller, D. (Princeton: Princeton University Press, 1995).

Goles, E., Schulz, O., and Markus, M., 'A biological generator of prime numbers', *Nonlinear Phenomena in Complex Systems* 3 (2000): 208.

Goodwin, B., *How the Leopard Changed Its Spots* (London: Weidenfeld & Nicolson, 1994).

Gorman, M., El-Hamdi, M., and Robbins, K. A., 'Experimental observation of ordered states of cellular flames', *Combustion Science and Technology* 98 (1994): 37.

Gould, S. J., *Wonderful Life* (London: Penguin, 1991).

Gould, S. J., *Ontogeny and Phylogeny* (Cambridge, Mass.: Belknap / Harvard University Press, 1977).

Gray, R. A., and Jalife, J., 'Spiral waves and the heart', *International Journal of Bifurcation and Chaos* 6 (1996): 415.

Gunning, B. E. S., 'The greening process in plastids. 1. The structure of the prolamellar body'. *Protoplasma* 60 (1965): 111.

Gunning, B. E., and Steer, M. W., *'Ultrastructure and the Biology of Plant Cells'* (Edward Arnold, 1975).

Gurdon, J. B., Harger, P., Mitchell, A., and Lemaire, P., 'Activin signalling and response to a morphogen gradient', *Nature* 371 (1994): 487.

Haeckel, E., *Art Forms in Nature* (Munich: Prestel, 1998).

Hardenberg, J. von, Meron, E., Shachak, M., and Zarmi, Y., 'Diversity of vegetation patterns and desertification', *Physical Review Letters* 87 (2001): 198101.

Harting, P., 'On the artificial production of some of the principal organic calcareous formations', *Quarterly Journal of the Microscopy Society* 12 (1872): 118.

Hassell, M. P., Comins, H. N., and May, R. M., 'Spatial structure and chaos in insect populations', *Nature* 353 (1991): 255.

Hayashi, T., and Carthew, R. W., 'Surface mechanics mediate pattern formation in the developing retina', *Nature* 431 (2004): 647.

Heaney, P., and Davis, A., 'Observation and origin of self-organized textures in agates', *Science* 269 (1995): 1562.

Henisch, H. K., *Crystals in Gels and Liesegang Rings* (Cambridge: Cambridge University Press, 1988).

Higgins, K., Hastings, A., Sarvela, J. N., and Botsford, L. W., 'Stochastic dynamics and deterministic skeletons: population behavior of the Dungeness crab', *Science* 276 (1997): 1431.

Hildebrandt, S., and Tromba, A., *The Parsimonious Universe* (New York: Springer, 1996).

Hwang, S.-M., Kim, T. Y., and Lee, K. J., 'Complex-periodic spiral waves in confluent cardiac cell cultures induced by localized inhomogeneities', *Proceedings of the National Academy of Sciences USA* 102 (2005): 10363.

Hyde, S., Andersson, S., Larsson, K., Blum, Z., Landh, T., Lidin, S., and Ninham, B., *The Language of Shape* (Amsterdam: Elsevier, 1997).

Hyde, S. T., O'Keeffe, M., and Proserpio, D. M., 'A short history of an elusive yet ubiquitous structure in chemistry, materials, and mathematics', *Angewandte Chemie International Edition* 47 (2008): 7996.

Jakubith, S., Rothemund, H. H., Engel, W., Oertzen, A. von, and Ertl, G., 'Spatiotemporal concentration patterns in a surface reaction: propagating and standing waves, rotating spirals, and turbulence', *Physical Review Letters* 65 (1990): 3013.

Kaminaga, A., Vanag, V. K., and Epstein, I. R., 'A reaction-diffusion memory device, *Angewandte Chemie International Edition* 45 (2006): 3087.

Kapral, R., and Showalter, K. (eds), *Chemical Waves and Patterns* (Dordrecht: Kluwer Academic, 1995).

Kareiva, P., 'Stability from variability', *Nature* 344 (1990): 111.

Kareiva, P., and Wennergren, U., 'Connecting landscape patterns to ecosystem and population processes', *Nature* 373 (1995): 299.

Kawczynski, A. L., and Legawiec, B., 'Two-dimensional model of a reaction-diffusion system as a typewriter', *Physical Review E* 64 (2001): 056202.

Kemp, M., *Visualizations* (Oxford: Oxford University Press, 2000).

Kemp, M., 'Divine proportion and the Holy Grail', *Nature* 428 (2004): 370.

Klausmeier, C. A., 'Regular and irregular patterns in semiarid vegetation', *Science* 284 (1999): 1826.

Koch, A. J., and Meinhardt, H., 'Biological pattern formation: from basic mechanisms to complex structures', *Reviews of Modern Physics* 66 (1994): 1481.

Kondo, S., and Asai, R., 'A reaction-diffusion wave on the skin of the marine angelfish *Pomacanthus*', *Nature* 376 (1995): 765.

Lawrence, P. A., *The Making of a Fly* (Oxford: Blackwell Scientific, 1992).

Lechleiter, J., Girard, S., Peralta, E., and Clapham, D., 'Spiral calcium wave propagation and annihilation in *Xenopus laevis* oocytes', *Science* 252 (1991): 123.

Lee, K.-J., McCormick, W. D., Pearson, J. E., and Swinney, H. L., 'Experimental observation of self-replicating spots in a reaction-diffusion system', *Nature* 369 (1994): 215.

Lemons, D., and McGinnis, W., 'Genomic evolution of Hox gene clusters', *Science* 313 (2006): 1918.

Li, C., Zhang, X., and Cao, Z., 'Triangular and Fibonacci number patterns driven by stress on core/shell microstructures', *Science* 309 (2005): 909.

Liaw, S. S., Yang, C. C., Liu, R. T., and Hong, J. T., 'Turing model for the patterns of lady beetles', *Physical Review E* 64 (2001): 041909.

Lipowsky, R., 'The conformation of membranes', *Nature* 349 (1991): 475.

Liu, R. T., Liaw, S. S., and Maini, P. K., 'Two-stage Turing model for generating pigment patterns on the leopard and the jaguar', *Physical Review E* 74 (2006): 011914.

Lotka, A. J., 'Analytical note on certain rhythmic relations in organic systems', *Proceedings of the National Academy of Sciences USA* 6 (1920): 410.

Lotka, A. J., 'Natural selection as a physical principle', *Proceedings of the National Academy of Sciences USA* 6 (1922): 151.

McKay, D., *et al.*, 'Search for past life on Mars: Possible relict biogenic activity in Martian meteorite ALH84001', *Science* 273 (1996): 924.

Mann, S., and Ozin, G., 'Synthesis of inorganic materials with complex form', *Nature* 382 (1996): 313.

Marder, M., Sharon, E., Smith. S., and Roman, B., 'Theory of edges of leaves', *Europhysics Letters* 62 (2003): 498.

Markus, M., and Hess, B., 'Isotropic cellular automaton for modelling excitable media', *Nature* 347 (1990): 56.

May, R., 'The chaotic rhythms of life', in Hall, N. (ed.), *Exploring Chaos. A Guide to the New Science of Disorder* (New York: W. W. Norton, 1991).

May, R. M., 'Simple mathematical models with very complicated dynamics', *Nature* 261 (1976): 459.

Mclean, R. J., and Pessoney, G. F., 'A large scale quasi-crystalline lamellar lattice in chloroplasts of the green alga *Zygnema*', *Journal of Cell Biology* 45 (1970): 522.

Meinhardt, H., *Models of Biological Pattern Formation* (London: Academic Press, 1982).

Meinhardt, H., 'Dynamics of stripe formation', *Nature* 376 (1995): 722.

Mertens, F., and Imbihl, R., 'Square chemical waves in the catalytic reaction $NO + H^2$ on a rhodium (110) surface', *Nature* 370 (1994): 124.

Michalet, X., and Bensimon, D., 'Vesicles of toroidal topology: observed morphology and shape transformations', *Journal de Physique II* 5 (1995): 263.

Michielsen, K., and Stavenga, D. G., 'Gyroid cuticular structures in butterfly wing scales: biological photonic crystals', *Journal of the Royal Society Interface* 5 (2008): 85.

Murray, J. D., 'How the leopard gets its spots', *Scientific American* 258(3) (1988): 62.

Murray, J. D., *Mathematical Biology* (Berlin: Springer, 1990).

Nijhout, H. F., *The Development and Evolution of Butterfly Wing Patterns* (Washington: Smithsonian Institution Press, 1991).

Nijhout, H. F., 'Polymorphic mimicry in *Papilio dardanus*: mosaic dominance, big effects, and origins', *Evolution and Development* 5 (2003): 579.

Nüsslein-Volhard, C., 'Gradients that organize embryo development', *Scientific American* (August 1996): 54.

Ortoleva, P. J., *Geochemical Self-Organization* (Oxford: Oxford University Press, 1994).

Ouyang, Q., and Swinney, H. L., 'Transition from a uniform state to hexagonal and striped Turing patterns', *Nature* 352 (1991): 610.

Ouyang, Q., and Swinney, H. L., 'Onset and beyond Turing pattern formation', in Kapral, R., and Showalter, K. (eds), *Chemical Waves and Patterns* (Dordrecht: Kluwer Academic, 1995).

Ozin, G. A., and Oliver, S., 'Skeletons in a beaker: synthetic hierarchical inorganic materials', *Advanced Materials* 7 (1995): 943.

Pearlman, H. G., and Ronney, P. D., 'Self-organized spiral and circular waves in premixed gas flames', *Journal of Chemical Physics* 101 (1994): 2632.

Pirk, C. W. W., Hepburn, H. R., Radloff, S. E., and Tautz, J., 'Honeybee combs: construction through a liquid equilibrium process?', *Naturwissenschaften* 91 (2004): 350.

Pratt, S. C., 'Gravity-dependent orientation of honeycomb cells', *Naturwissenschaften* 87 (2000): 33.

Prost, J., and Rondelez, F., 'Structures in colloidal physical chemistry', *Nature* 350 (supplement) (1991): 11.

Richardson, M. K., and Kueck, G., 'A question of intent: when is "schematic" illustration a fraud?', *Nature* 410 (2001): 144.

Saunders, P. T. (ed.), *Morphogenesis: Collected Works of A. M. Turing*, vol. 3 (Amsterdam: North-Holland, 1992).

Schopf, W. (ed.), *Earth's Earliest Biosphere* (Princeton: Princeton University Press, 1991).

Scott, S., 'Clocks and chaos in chemistry', in Hall, N. (ed.), *Exploring Chaos. A Guide to the New Science of Disorder* (New York: W. W. Norton, 1991).

Scott, S. K., *Oscillatios, Waves, and Chaos in Chemical Kinetics* (Oxford: Oxford University Press, 1994).

Sharon, E., Marder, M., and Swinney, H. L., 'Leaves, flowers and garbage bags: making waves', *American Scientist* 92 (2004): 254.

Shipman, P., and Newell, A. C., 'Phyllotactic patterns on plants', *Physical Review Letters* 92 (2004): 168102.

Sick, S., Reinker, S., Timmer, J., and Schlake, T., 'WNT and DKK determine hair follicle spacing through a reaction-diffusion mechanism', *Science* 314 (2006): 1447.

Smolin, L., 'Galactic disks as reaction-diffusion systems', preprint <http://www.arxiv.org/abs/astro-ph/9612033> (1996).

Solé, R., and Goodwin, B., *Signs of Life* (New York: Basic Books, 2000).

Steinberg, B. E., Glass, L., Shrier, A., and Bub, G., 'The role of heterogeneities and intercellular coupling in wave propagation in cardiac tissue', *Philosophical Transactions of the Royal Society A* 364 (2006): 1299.

Stevens, P. S., *Patterns in Nature* (London: Penguin, 1974).

Stewart, I., *Nature's Numbers* (London: Weidenfeld & Nicolson, 1995).

Stewart, I., *Life's Other Secret. The New Mathematics of the Living World* (New York: Wiley, 1998).

Stewart, I., *What Shape is a Snowflake? Magical Numbers in Nature* (London: Weidenfeld & Nicolson, 2001).

Stewart, I., and Golubitsky, M., *Fearful Symmetry* (London: Penguin, 1993).

Suzuki, N., Hirata, M., and Kondo, S., 'Travelling stripes on the skin of a mutant mouse', *Proceedings of the National Academy of Sciences USA* 100 (2003): 9680.

Swinton, J., 'Watching the daisies grow: Turing and Fibonacci phyllotaxis', in Teuscher, C. A. (ed.), *Alan Turing: Life and Legacy of a Great Thinker* (Berlin: Springer, 2004): 477.

Theraulaz, G., Bonabeau, E., and Deneubourg, J.-L., 'The origin of nest complexity in social insects', *Complexity* 3 (1998): 15.

Thomas, E. L., Anderson, D. M., Henkee, C. S., and Hoffman, D. 'Periodic area-minimizing surfaces in block copolymers', *Nature* 334 (1988): 598.

Thompson, D'Arcy W., *On Growth and Form* (New York: Dover, 1992).

Turing, A., 'The chemical basis of morphogenesis', *Philosophical Transactions of the Royal Society B* 237 (1952): 37.

Vilar, J. M. G., Solé, R. V., and Rubí, J. M., 'On the origin of plankton patchiness', *Physica A* 317 (2003), 239.

Vincent, S., and Perrimon, N., 'Fishing for morphogens', *Nature* 411 (2001): 533.

Vogel, G., 'Auxin begins to give up its secrets', *Science* 313 (2006): 1230.

Waldrop, M. M., *Complexity* (London: Penguin, 1992).

Weaire, D., and Phelan, R., 'Optimal design of honeycombs', *Nature* 367 (1994): 123.

Weaire, D., and Phelan, R., *The Kelvin Problem* (London: Taylor & Francis, 1996).

Weaire, D., and Phelan, R., 'The structure of monodisperse foam', *Philosophical Magazine Letters* 70 (1994a): 345.

Weaire, D., Phelan, R., and Phelan, R., 'A counter-example to Kelvin's conjecture on minimal surfaces', *Philosophical Magazine Letters* 69 (1994b): 107.

Weyl, H., *Symmetry* (Princeton: Princeton University Press, 1969).

Whitfield, J., *In The Beat of a Heart: Life, Energy, and the Unity of Nature* (Washington, DC: Joseph Henry Press, 2006).

Whittaker, R. H., *Communities and Ecosystems*, 2nd edn (New York: Macmillan, 1975).

Winfree, A. T., *When Time Breaks Down* (Princeton: Princeton University Press, 1987).

Winter, A. and Siesser, W. G. (eds), *Coccolithophores* (Cambridge: Cambridge University Press, 1994).

Wintz, W., Dobereiner, H. G., and Seifert, U., 'Starfish vesicles', *Europhysics Letters* 33 (1996): 403.

Zhu, A. J., and Scott, M. P., 'Incredible journey: how do developmental signals travel through tissue?', *Genes & Development* 18 (2004): 2985.